A LITTLE ABOUT SURFACE MOUNT TECHNOLOGY

Printed Circuit Board Assembly Process

Adibhatla Krishna Rao

Copyright © Adibhatla Krishna Rao 2025
All Rights Reserved.

ISBN
Paperback 979-8-89906-734-1
Hardcase 979-8-89906-735-8

This book has been self-published with all reasonable efforts taken to make the material error-free by the author. No part of this book shall be used, reproduced in any manner whatsoever without written permission from the author, except in the case of brief quotations embodied in critical articles and reviews.

The Author of this book is solely responsible and liable for its content including but not limited to the views, representations, descriptions, statements, information, opinions and references ["Content"]. The Content of this book shall not constitute or be construed or deemed to reflect the opinion or expression of the Publisher or Editor. Neither the Publisher nor Editor endorse or approve the Content of this book or guarantee the reliability, accuracy or completeness of the Content published herein and do not make any representations or warranties of any kind, express or implied, including but not limited to the implied warranties of merchantability, fitness for a particular purpose. The Publisher and Editor shall not be liable whatsoever for any errors, omissions, whether such errors or omissions result from negligence, accident, or any other cause or claims for loss or damages of any kind, including without limitation, indirect or consequential loss or damage arising out of use, inability to use, or about the reliability, accuracy or sufficiency of the information contained in this book.

Made with ♥ on the Notion Press Platform

www.notionpress.com

CONTENTS

PREFACE

This technical document presents a qualitative description of the electronic manufacturing industries and various practices adopted to meet their product quality standards. This includes detailed descriptions of manufacturing processes and the manufacturing enterprise that will help readers of this book to know about various electronic manufacturing industries, the demand for electronic products, and global business requirements. It provides a complete idea about the electronic manufacturing process and important concepts in detail and comes to know "A little about everything."

This book presents technical information for students of engineering at a post-graduate level about basic knowledge on printed circuit boards (PCB), semiconductors, automation, and processes adopted in manufacturing industries. The author tried to elaborate on his work experience with automated machines, production flow, critical processes, and assembly process flow to provide up-to-date technology that provides a solid background on PCB assembly processes to face new challenges in this digital world. A sustained effort has been made to make the reader's clear understanding through relevant pictures, with an objective "Knowledge Sharing Program"

This Volume 1 book mainly covers the front-end manufacturing processes – scope of electronics, manufacturing Industry briefly, A little about bear board PCB, printed circuit board, semiconductor packages, PCB assembly Manufacturing process briefly, Screen printing Process, Solder paste Inspection SPI, component placement process, and reflow process.

Every effort has been made to elaborate topics for a layman's understanding with pictures, equations, and shared in-depth knowledge combined with personal experience while dealing with automated machines, included in a brief explanation on day-to-day challenges that arose on the manufacturing floor and at customer service.

I hope you find this book helpful in rounding out PCB assembly knowledge, You need to become a successful employee, an engineer, a production supervisor, and a relatively stress-free Manager. Automation equipment suppliers, sales, and after-sale service persons also benefit while dealing with their customers and understanding for improvements.

ACKNOWLEDGEMENTS

In preparing this article, some content material is absorbed from various articles, white papers, and technical documents by SMT professionals and design engineers of OEMs. Some images are absorbed from the Google App only to make it easier to understand technical readers. By any chance, if the images are subjected to Copyright, the author is urging to forgive for the same.

Special thanks to Mr. Soni Saran Singh, CEO, NMTronics India Pvt Ltd., who gave me enough opportunities to deal with customers of different production environments and encouraged me to conduct training programs for my Manufacturing Industry customers' in-house service engineers to enhance their process and machine knowledge. This, in turn, prompted me to write this book for your benefit.

Special thanks to my Principle key players Fuji corporation Japan (SMT pick and Place machines, Printers, Robotic solutions), ITW Inc. USA (Screen Printers, Reflow Machines, Glue Dispensing Machines), Koh Young technology Inc Korea (Solder Paste Inspection, Automated Optical Inspection), Heller Usa (Reflow oven solutions) of SMT technology (OEMs). Relevant stuffings absorbed from these key players and associated ancillary product manufacturers are needed for this book presentation to make the next generation ready for the new upcoming technologies.

Regardless of the source, I would like to thank all my engineers who have provided lots of information and clarified technical doubts about SMT Technology. I doubt I could have completed this book without their support.

Finally, I would like to give endless special thanks to my wife, Adibhatla Usha, who encouraged me a lot, gave me hope in stressful situations, and always tells me never to give up.

It is a well-known fact and well-accepted argument that electronics have become a part of human life. In the absence of electronics, it is very difficult to imagine where our lives would be. Since the late 1990s, drastic developments have happened in the electronics industry and played a significant role in modern life, from the smallest consumer electronics to the control systems of the largest automated production lines. The introduction of powerful hardware, Software, various types of electronic circuits, and embedded systems are becoming fundamentals of large applications.

These include manufacturing, Information and Technology, consumer electronics, Automotive Industry, Aerospace communications, Defence, Medical, Energy, Safety Engineering, lightning technology, and the telecom industry. In addition to these, many other rising sectors like Satellite systems, Robotics, semiconductors, Solar systems, and Industrial atomization are playing vital roles in the advancement of technology. Even in the latest intelligent buildings, electronic control devices are used to automate appliances and manage the Environment.

With the development of devices such as smartphones and GPS satellite navigation systems and developments in the medical field such as medical imaging, patient diagnosis, and care delivery with remote monitoring, twenty-first–century technology is going to maintain its current rate of advancement to meet supply demands and produce more feature-rich and powerful miniature devices, especially in consumer electronics.

INTRODUCTION

The role of electronics and its developments in various sectors can be categorized phase-wise – mid-80s, late 90s, and twenty-first century. The development of miniaturized electronic products and growing demands evoked electronic manufacturers to make changes or requirements to changes in their manufacturing process rules and regulations, process methods, production concepts, and revolution in quality management.

Phase-by-phase changes are happening in circuit design, component miniaturization, and design, especially developments in semiconductors in terms of design for manufacturability. This made equipment manufacturers and their ancillary co-product developers bound to come up with new model equipment to meet demands in terms of speed, reliability, and capability, which in turn made a revolution in Surface Mount Technology.

So, twenty-first-century technology is likely to maintain its current rate of advancement with noticeable progression in various fields- we can say every field. Manufacturing companies, product designers, product developers, and Researchers will need to stay up to speed and make sure that they have the current relevant connections in the industry. Here, we are going to re-touch electronics in various technical fields with some examples that are influencing our day-to-day life. Electronics plays a role in every sector, and it gives immense ideas about the technology and importance of engineering.

ROLE OF ELECTRONICS IN THE AUTOMOBILE INDUSTRY

In automotive systems, more and more equipment is being changed from mechanical systems to electronic systems. Because of its versatility and flexibility, the embedded system has become the heart of a vehicle's electronic system. An embedded system is a combination of computer hardware and software designed for specific logical functions such as an embedded airbag system, navigation system, adaptive cruise control system, embedded rain sensing system, automatic parking system, etc. Similarly, electronically controlled functions like fuel injection control devices, transmission controls, entertainment systems, lights, horns, wipers, defrosters, etc., and sensor controls are all witnessed in present luxury cars.

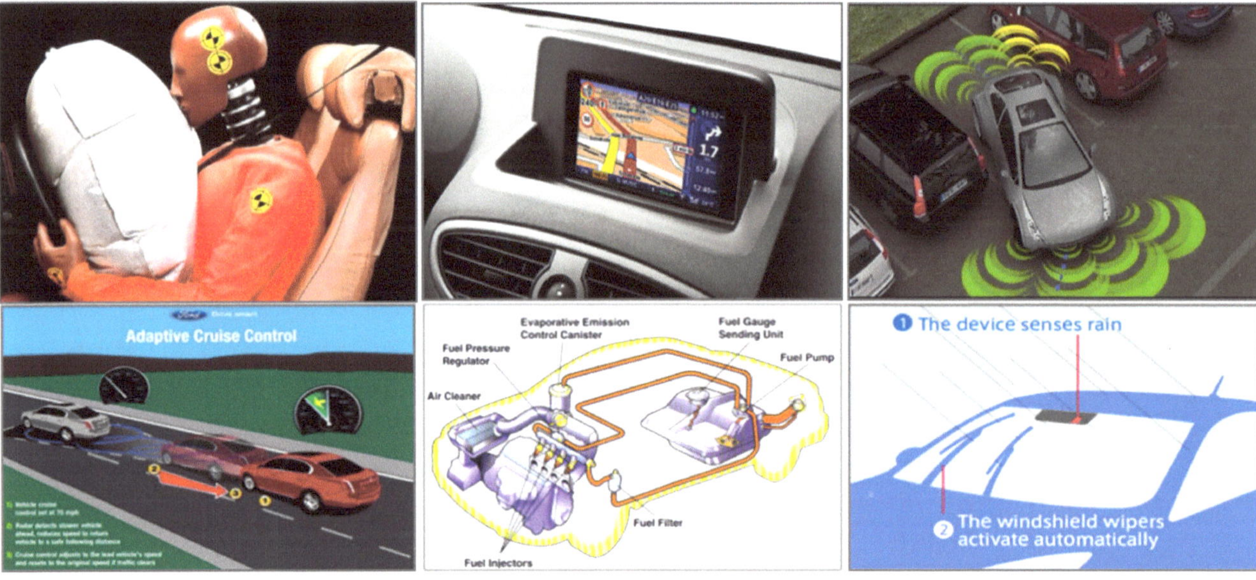

Developments in fuel injection control devices, fuel combustion, power train crash protections, and the advanced usage of embedded systems in vehicles can help control pollution. It is the revolution of electronics that has manipulated automotive design, safety, and features to consumer demand. Today, a typical vehicle contains around 25 to 35 microcontrollers, and some luxury vehicles contain approximately 60 to 70 microcontrollers per vehicle. It is a transformation from mechanical systems to electronics.

CONSUMER ELECTRONICS

Consumer electronics that are intended to be used in daily life. Just compare any home Computer or mobile phone you are using today with one you had back in the 1980s and it is obvious how things have changed. Even smartphones, GPS navigation, and digital trading are considered consumable items. In education, consumer devices gained popularity in representing the contents in graphical, 3D views and more adoptable methods for clear understanding and easy learning.

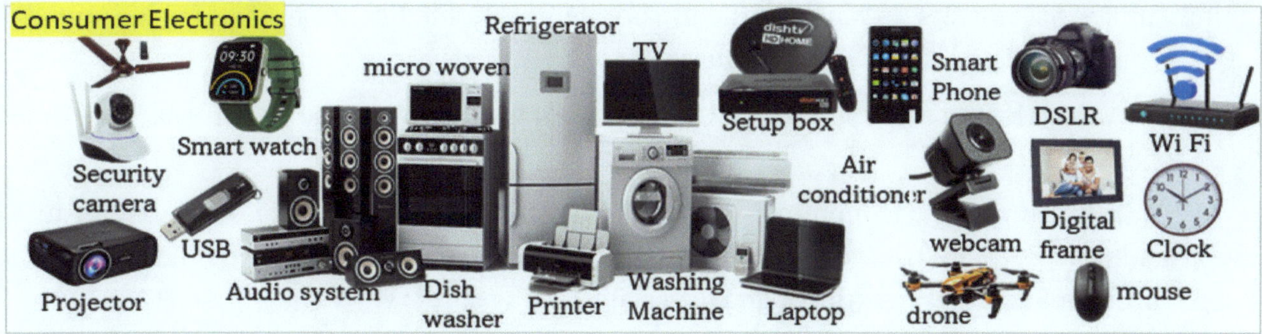

Our day starts with smart smartphones and laptops, which are now a part of our lives. The style of our communication drastically changed over the decades.

ROLE OF ELECTRONICS IN DEFENCE

Satellite communication, radar, instrumentation, wireless communication, and intelligent tracking systems play a major role in the defense sector, and they always opt for special design and manufacturing with stringent manufacturing processes and methods. Products include RF Data links, Search and rescue avionics, customized radio products, High-Speed Video and Data Transmissions, and SONAR, which is a modern technology that helps us to track submarines, fish, shipwrecks, map the seabed, and other Navigational purposes.

All Défense products are mostly considered Class 3 products, for which manufacturing industries never compromise on quality. Manufacturing process instructions strictly adhere to the given specifications. Examples include electronic warfare systems, air-naval ground-based applications, radar ponders, acoustic communication systems, and high-power transmitters.

Oil and Gas exploration, Agriculture, mining, global navigation, and satellite-based systems are the best examples. Industries related to these fields manufacture and design specifically for the intended use.

ROLE OF ELECTRONICS IN MEDICAL TECHNOLOGY

The use of electronics in the medical sector improved the quality of diagnosis and medical treatment. Electronic manufacturing for medical equipment such as diagnostic tools, medical imaging, and monitoring systems. Similarly, devices used for electronic medical records and medical electronics. Developments in optics technology, and Vision Engineering influenced the role of electronics in medical electronics through the invention of sophisticated precision medical equipment so that doctors and surgeons can confidently do medical examinations and modify medical treatment. Let us count some examples ...

MRI - Magnetic Resonance Imaging - Used in radiology, to review the natural object of inner elements of the body.

ECM –Electronic cardiac monitor -Used to display the electrical and pressure waveforms of the cardiac system. ECG observes irregular activity within the cardiac system and heart issues. It is used throughout medical treatment and especially during surgery.

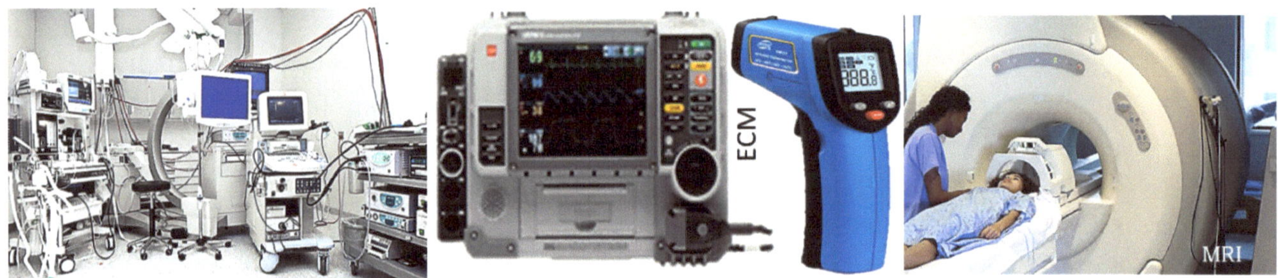

Defibrillator - A defibrillator is used in emergency conditions, like a heart attack occurring

IR Thermometers Measure temperature by detecting infrared radiation generated by the body. These devices are often used in airports to detect the condition of passengers' health from a distance and observe diseases like viral fever, such as EBOLA, SAARC, etc.

Brain Wave Machine - one type of instrument in medical electronics that is used to record the electrical activity of the scalp with Electroencephalography by the firing of neurons within the brain

A blood Gas analyzer is an instrument that is used to calculate the pressure of the chemical elements like carbon monoxide, nitrogen, and oxygen in the blood.

INDUSTRIAL ELECTRONICS

Industrial electronics is a branch of electronics that deals with power electronic devices such as thyristors, SCRs, AC/DC drives, meters, sensors, analyzers automatic test equipment, multimeters, data recorders, relays, resistors, semiconductors, waveguides, scopes, amplifiers, radio frequency (RF) circuit boards, timers, counters, etc.

It covers all the methods and facets of control systems, instrumentation, mechanism and diagnosis, signal processing, and automation of various industrial applications. The core research areas of industrial electronics include electrical power machine designs, power conditioning, and power semiconductor devices. So, to put it simply, industrial electronics refers to equipment, tools, and processes that involve electrical equipment in an industrial setting. This could be a laboratory, automotive plant, power plant construction site, etc. Industrial electronics are also used

extensively in chemical processing plants, oil/gas/petroleum plants, mining and metal processing units, electronics, and semiconductor manufacturing.

ROLE OF ELECTRONICS IN INSTRUMENTATION

Test and measuring devices, process control type of equipment, signal-generating devices, industrial control systems, etc., come under this category.

Test and Measurement Devices				Process Control Equipment ↓	Signal Generating Devices ↓	Industrial Control Systems ↓
Measuring instruments ↓	Circuits Under Test ↓	Response circuit Analyzer ↓	Analyzers ↓	Logic Controllers PLC Systems	Signal generator	Circuit protect devices
Volt Meters	Power Supplies	Oscillo-scopes	Logic Analyzer	Count down Timers	Function generator	Electrical instruments
Ammeters	Signal Generators	Frequency Counter	Spectrum Analyzer	Power Controllers	Pulse generator	Temperature Control Devices
Galvano Meters	Pulse Generators		Protocol Analyzer	Digital Temp Indicators	Frequency synthesizer	
Multi Meters	Pattern Generators			Digital Controllers		

ROLE OF ELECTRONICS IN LIGHTING TECHNOLOGY

For decades, Tungsten Bulbs and Fluorescent tube lights ruled the golden era of lighting technology. Stage lighting, like flood lights, light control, moving heads, laser technology, flight cases, and rock mounts, had improved to some extent, but it was limited to entertainment purposes and did not focus on Energy efficiency.

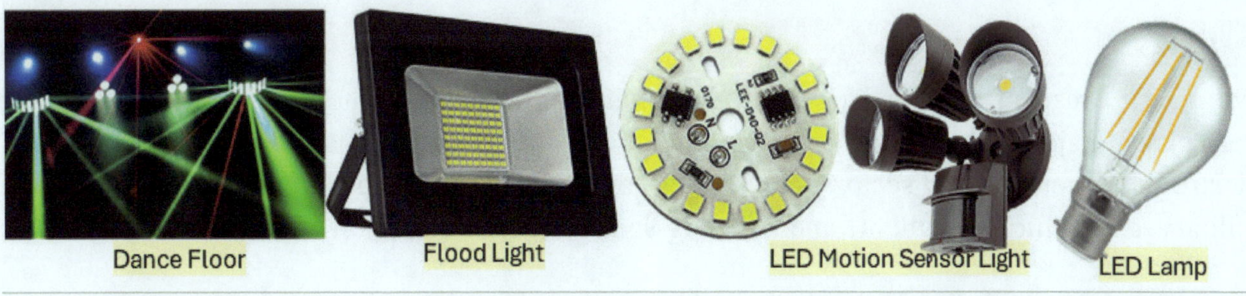

Dance Floor   Flood Light   LED Motion Sensor Light   LED Lamp

Now, in the early 21st century, an era of smart lighting has started. Designers and inventors played a significant role in developing lighting technology with the aim of energy efficiency and achieving some aesthetic or practical effect. Smart Lighting is a lighting technology designed for energy efficiency that enables minimizing and saving light by allowing the householder to control remotely cooling and heating, lighting, and the control of appliances.

Here are some examples of electronic items – motion detectors, electronic control units, and Controllable switch relays that drive lighting applications, especially for sensor types passive, Infrared, ultra-sonic, and hybrid. Rapid and frequent switching, control of illumination, and photoelectric controls are also added examples of inventions.

Great applications in all fields, just because of its high efficiency with minute energy consumption. LED Bulbs and LED TVs are just common examples to accept the fact of how LEDs drive the electronic manufacturing industry.

ROLE OF ELECTRONICS IN ENERGY

Power electronics played a significant role in the interfaces between the Distributed Generation (DG) sources such as solar energy and wind energy. For these power systems, it is necessary to add control loops. i.e., energies are converted to electrical energy using the power electronic converters. The technology of power electronics converters has evolved dramatically based on the semiconductor's advancement.

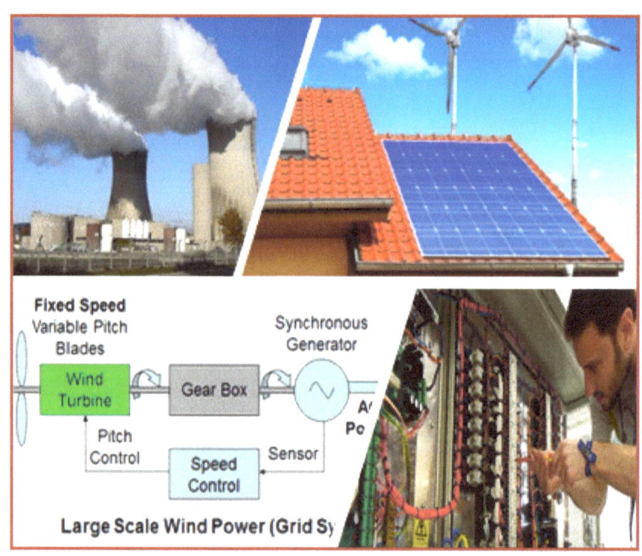

Power Electronics allows solar energy to be used by converting the direct current produced by solar panels into AC used in commercial electrical grids. Wind energy also needs to be converted and must be fed into a grid at a constant frequency despite changing wind conditions.

Other forms of alternate power such as Thermal, Hydro, and Nuclear also take advantage of power electronics for effectively delivering power. Everything that has power steering in your car, microwave ovens, and battery chargers utilize the power electronics. Almost every industry uses power electronics, including desalination plants, painting vehicles, or testing rail cars.

CONCLUSION

We need to respect the Importance of electronics and their applications in today's world, as they play a major role in our lifestyle, and it is very difficult to go back to the past. Growth of electronics observed in all sectors –modern electronics, power, energy, Medical, Aviation, space, military electronic equipment, and systems such as communication devices, radar systems,

weapon systems, and communications, and there is a tremendous increase in electronic manufacturing company establishments to meet the supplies.

Technology keeps on changing to make products more energy-conservative, miniature, precise, and with more space-age capabilities. One thing is sure: PCB assembly or circuit assembly is a MUST for all the above.

So, any electronic product contains PCB assembly or assemblies as a part of the complete product, and the complexity of those PCB assemblies depends on the robust circuit design. Look at any PCB assembly, assembly of electronic circuit products, assembly involves surface mount components, through-hole components, electrical components, mechanical components, wire harnessing, or even plastic components interconnected with PCB through the soldering method. The design of any PCB assembly is linked with the product design for its intended input and desired output.

Also, wherever logic is required in a product, an electronic circuit is required. Hence, PCB assembly is required. Hence, there is a tremendous demand for electronic manufacturing units all over the world, and global PCB assembly manufacturers are constantly pushing the boundaries of mass production processes to produce various assembly products to meet the ever-increasing demands of their users and customers.

Conventional production processes cannot help manufacturers meet the demands and produce quality, reliable, high-performance products due to miniaturization. The process of making miniaturized electronic products smaller, lighter, and more powerful. Miniaturization made electronics more widely used in many fields.

PRODUCT MINIATURIZATION

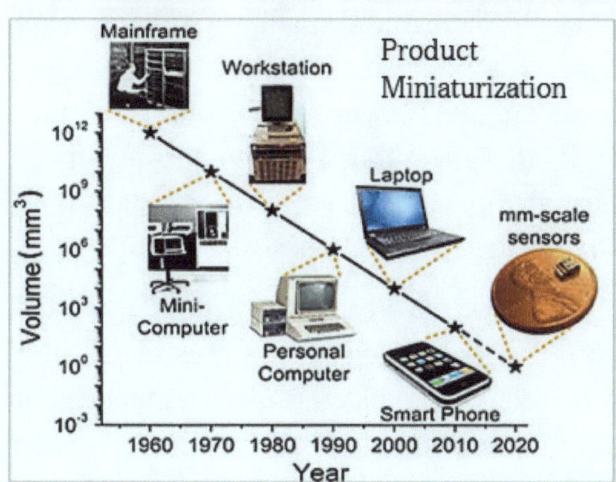

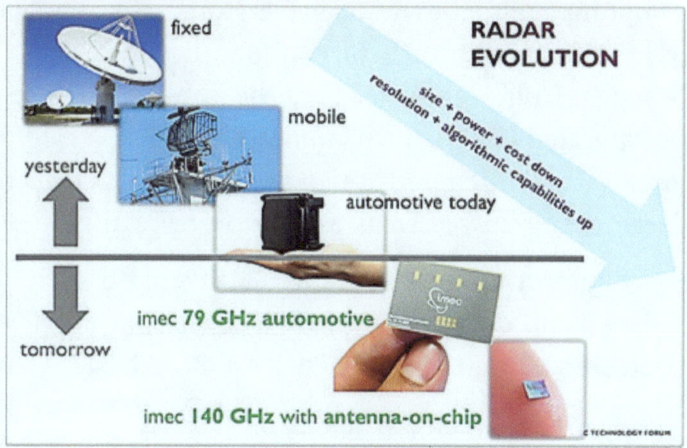

Industries that demand greater speed, efficiency, power, and decreased weight are pushing electronics to miniaturize. Shrinking electronics while maintaining their function requires massive amounts of R&D, while the focus on cost is put on the back burner. Processor dyes have been consistently shrinking for years, from 90nm 10 years ago to 14 nm and less than one nanometer today. PC board fabrication has witnessed a 75 -85 % reduction in size.

APPLICATION MINIATURIZATION

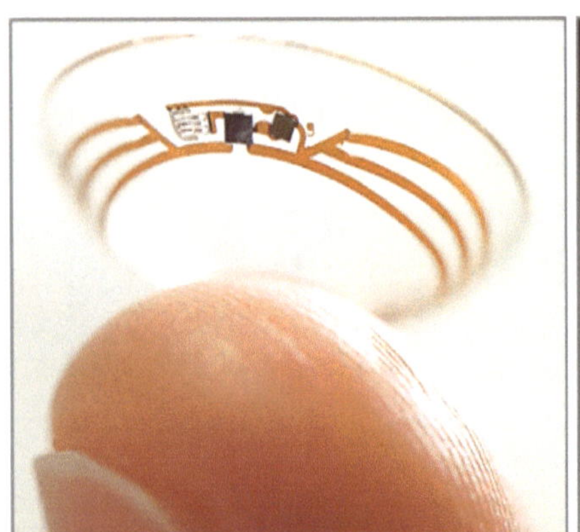
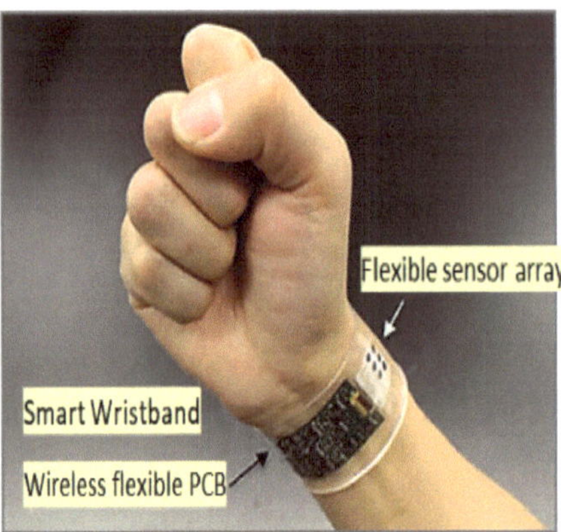

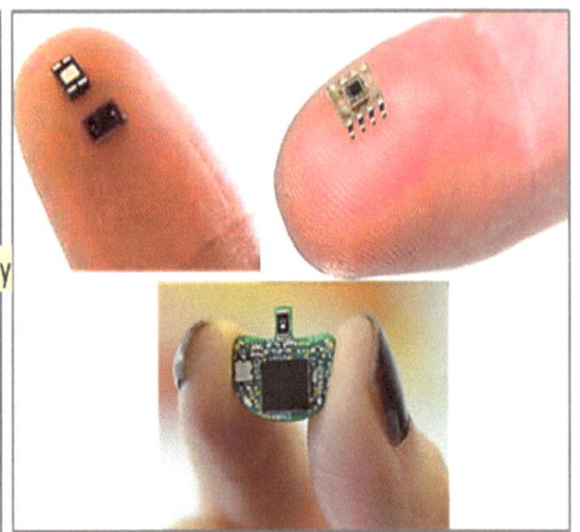

COMPONENT MINIATURIZATION

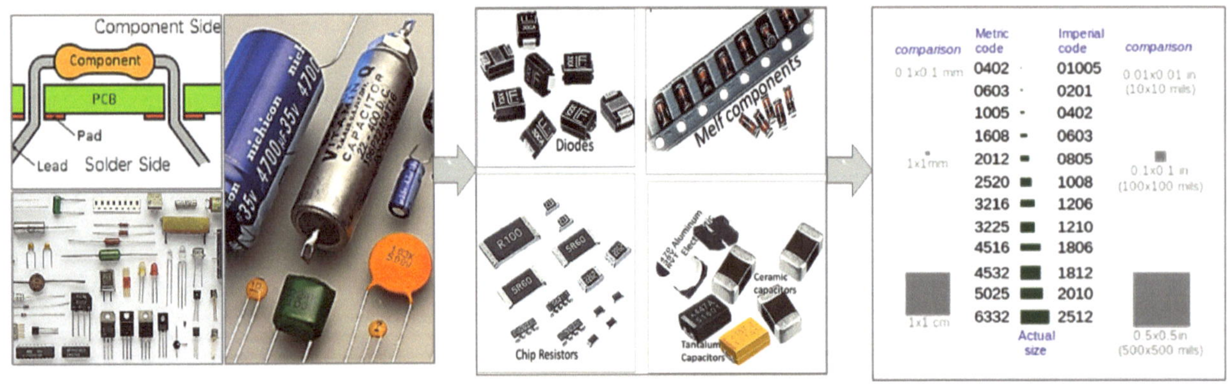

ACTIVE COMPONENT MINIATURIZATION

Component miniaturization in electronics is an endless march. Smaller and smaller – Never ending cycle. Miniaturization will continue if consumers demand product innovations. Components disappear from your naked eye. The invention of Integrated circuits ICs with billions of transistors incorporated less than a nanometer in size. Every effort by designers to make products as tiny as possible, with high performance, with endless features, and applications. Accordingly, designing electronic circuits as small as possible using high-density packages integrates advanced devices.

Engineer's Electronic Circuit Invention of a product, technically transformed into PCB design, to produce the electronic product or application. Product developers connect Precision Electronic Circuit Components on Board base professionally and electrically connect them through conductive traces and pads using advanced soldering technology. The device used in the

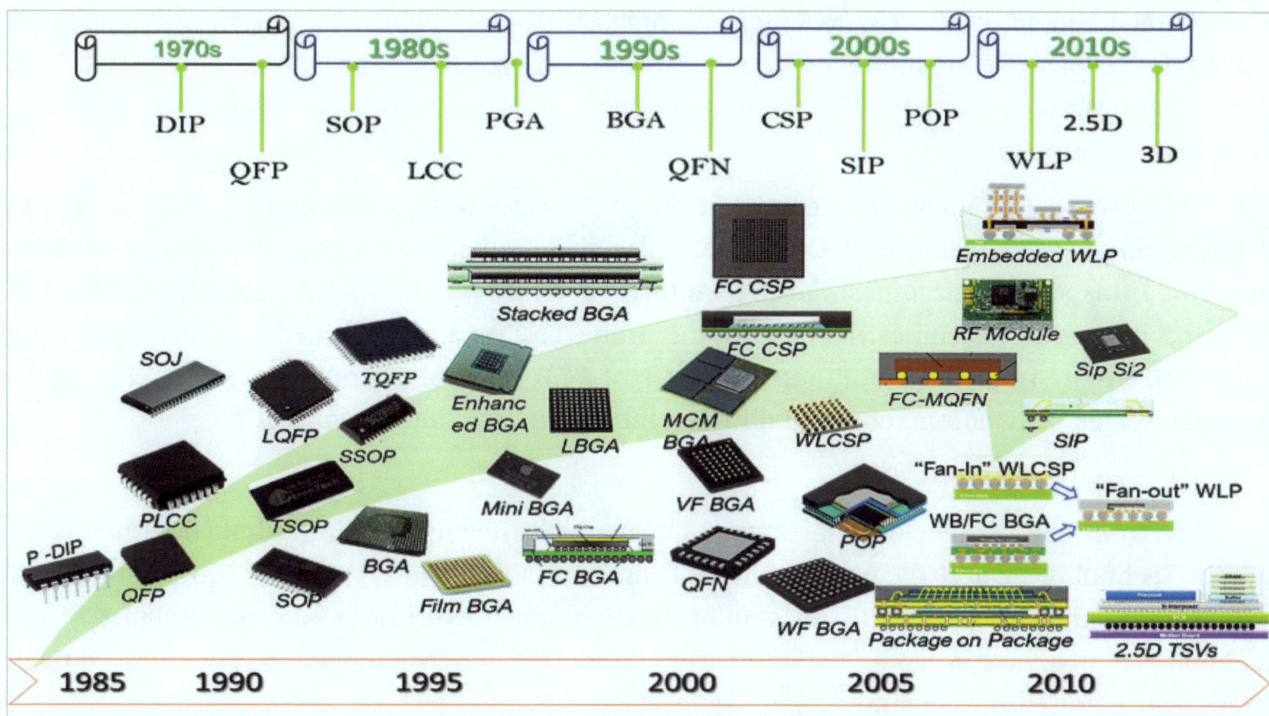

circuit includes Input /Output connections, I/O ports, and power supply connections. Engineering Circuit development, transformed into a technical way of PCB design considering Traces (wires of PCB), High current traces, signal traces, power traces, Ground traces, required PCB size (multi-layers), and suitable for Manufacturing process.

SURFACE MOUNT TECHNOLOGY – PCB ASSEMBLY PROCESS

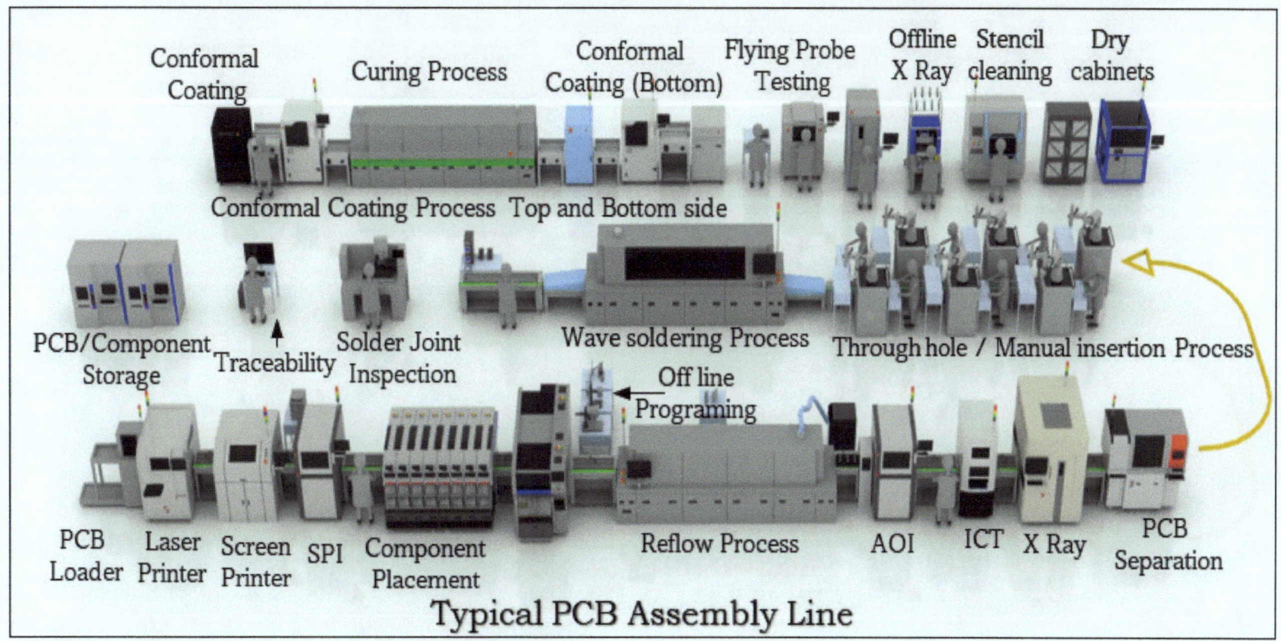

Typical PCB Assembly Line

Electronic manufacturing PCB assembly automated production lines consist of a series of workstations or several automated machines linked together through an automated PCB conveying system to move the PCB assembly between the specific process stages since assembly is constructed step-wise.

This process assembly is a combination of Process related machines (such as Screen printers, reflow, Wave soldering), Component placement machines (such as pick and Place, Auto insertion), Inspection machines like SPI, AOI, X-ray and testing types of equipment like and Test and measuring machines like ICT, flying probe Testers and dedicated test and measuring solutions. Apart from these, different varieties of PCB assembly handling /carrier equipment between stages, to handle assembly and to link the production line equipment.

The front end of any electronic manufacturing or PCB assembly line is linked with Surface mount Technology, where the production SMT line consists of process types of equipment, board handling, component placement, Inspection, testing, and rework types of equipments. Prime technologies involved in SMT are Automation, Soldering, Thermodynamics, robotics, Optical, X-ray, Laser, and Plasma technology.

SUMMARY

Where there is a logic in an application, the presence of an electronic product or application must be there. To meet demands, manufacturers must adopt new technologies. Incorporating Precision industrial robots, automation, in factories is mandatory to meet global volumes. Smart factory Concept, sensor-based inspection, package advancement, design for manufacturability, and advanced level in digitization are vital for the next generation electronics. Devices, Internet of things, smart sensors, digitization, Artificial intelligence AI, and big data are going to change the future of Electronic Manufacturing

One business drives the other business. Here, in our scenario, Telecom manufacturers, EMS (electronic manufacturing services), and Contract manufacturers, automotive electronics manufacturers, Govt organizations, specialized electronics manufacturers. etc., are all prime business holders. Business effective operation, growth, revenue, success, heavily reliant on the performance of secondary businesses such as semiconductor, supply chain, product design and development, training, standards, and others, as shown in Fig. Business expanding globally with a compound annual growth rate.

Opportunities are expanding the careers in certain trades

PREFACE

Electronic packaging is a multidisciplinary subject. It combines mechanical, Electrical, Electronic, Industrial engineering, Physics, Chemistry, and even Marketing. The commonly accepted approach of standard hierarchy levels of semiconductors is Level 0, Level 1, Level 2, Level 3, and Level 4 for electronic Interconnections.

Level 0 means Gate-to-Gate Interconnection on a silicon Die. The next step of electronic interconnection is Level 1, where a silicon die is bonded to the package through a wire bonding process. In the level 2 process, the PCB is from Component to Component or external connector. Further levels can be connections between PCBs and Subassemblies. The electronic industry continues to explore and develop new methods to fabricate and interconnect its Electrical and electronic devices. Unfortunately, these hierarchy levels no longer work for today's more complex approaches toward the interconnection of electronic elements.

The present evolving hierarchy of below Interconnection Levels completely changed its terminology and is very difficult to categorize.

- Wafers to Wafer Stacking

- Chip-to-wafer level packaging

- Single chip package to Stacked chip (POP, PIP, PUP) or multi-chip packages (e.g., System in package)

- Daughter board assembly/sub-assembly to embedded systems

- Motherboard assembly to stacked embedded systems.

The backend process of semiconductors involves material science technology, mechanical design, Electrical layout, modeling, and many other engineering specialties. For better understanding and awareness, here is the backend process of a semiconductor package. Individual discrete components are typically etched in silicon wafers before being cut and assembled in a package. The casing on this assembly can be metal, plastic, glass, or ceramic. It protects against impact and corrosion, which holds the contact pins or leads that are used to connect from external circuits to the device, and dissipates heat produced in the device.

Our topic especially addresses Semiconductor packages and focuses on providing a fundamental understanding of component Identification, component mounting techniques, Package dimensions, package classification, and other engineering properties that are essential for PCB assembly production.

ELECTRONIC PACKAGING

An electronic product designed to perform its intended function, the final box-build product will integrate mechanical components, electrical components, electronic components, plastic molds, sub-assemblies, etc., or electronic components to form a single component, an assembly, or a desired product. It offers major functions like.

- Interconnection of electrical signals

- Mechanical protection of the circuit

- Distribution of power (electrical energy) for circuit functions

- Dissipation of heat generated by the circuit function

- Manufacturability and serviceability

Any organization or designer who wants to develop new electronic hardware for a particular or dedicated application must follow a set of startup processes, such as Preliminary Production design, schematic diagram, PCB layout, Final BOM (Bill of material), prototype, and finally, Test and verification. This means product development is a combination of electronic design, mechanical design, and software design.

When it comes to Electronic Design, the Design is in a Schematic diagram, i.e., an electrical circuit. So, the circuit is an integration of electronic components, Electrical components, Mechanical components, and external harnessing. The selection of component category purely depends on the hardware requirement for that circuit design.

On successful acceptance of schematic design, followed by Prototype testing, circuit design transforms to physical form for product development. Perfectly designing PCB layout and selection of semiconductor Components is key to small-scale production, cost optimization, and procurement chain support for the product performance.

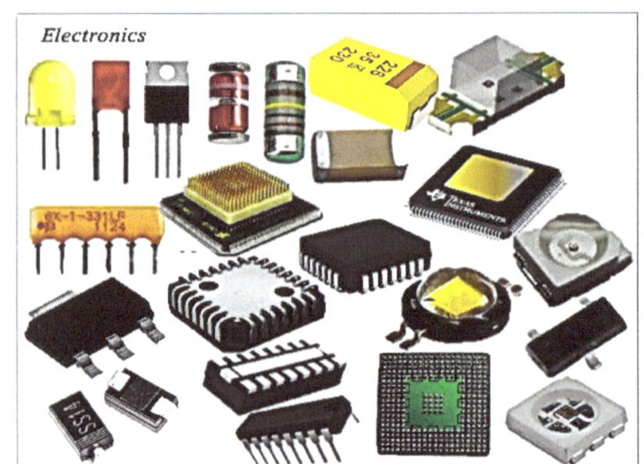

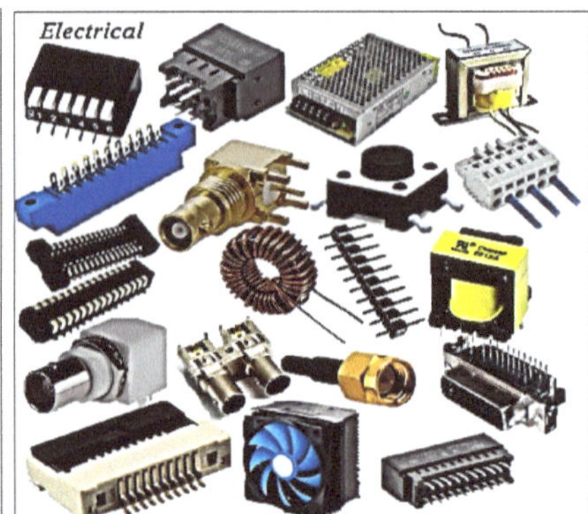

The designer considers so many parameters while selecting the required electronic packages for various applications like high-speed communications, high-frequency data links, fiber optic lines, Industrial applications, etc., where components capable of efficient function in innumerable environments such as surge handling, environmental stress, moisture resistance, thermal shocks, etc. Some high-grade components are specially designed for Aerospace, Avionics applications, automotive, and military applications. Hence, the range of component selection purely depends on the desired performance and efficiency of the product-designed application. In other words, we can say it is product life cycle management.

COMPONENT CLASSIFICATIONS

As explained earlier, electronic assembly integrates semiconductors, electronic components, electrical components, mechanical and external wire harnessing, and plastic parts. All these chosen components are mounted on a schematically designed PCB printed circuit board and electrically connected through conductive traces on the PCB.

For our clear understanding of their functionality, apart from mechanical components, electronic components are classified into Passive and Active components. Passive components are those that do not require energy or a power source to operate, except for available AC circuits that are connected to them. i.e., not capable of power gain and is not a source of Energy. Resistors, capacitors, inductors, Transformers, switches, and some diodes are examples of electronic Passive components. Whereas Active components require external and conditional sources to operate in the circuit. Oscillator transistors and small and large-scale Integrated circuits (ICs) are some examples that come under this category. Active components can amplify signal or power gain.

Regardless of Passive, active, or even mechanical components available in the Global market in two classifications for mounting purposes on the PCB assembly. 1 Through hole components (THC), and 2. Surface mount devices (SMDs). Same electronic component with the same functionality, most Electronics, electrical, and mechanical components are available in both THC and SMD form. Each classification electronic package has its advantages and disadvantages.

Semiconductors are becoming an essential part of the high-tech world. Semiconductor companies across the globe are coming up with technological advancements. Innovative semiconductor applications are driving towards prompting problem-solving through vast changes in packaging. Because of new semiconductor applications, the effectiveness of electronic systems, reliability, and cost, the product developer /designer chooses the applicable package for their circuit assembly. At the same time, the opted components or packages must withstand production process requirements because components are bonded through conductive traces on PCB through a soldering process in high temperatures ranging from 220°C - 300°C.

THROUGH HOLE COMPONENTS

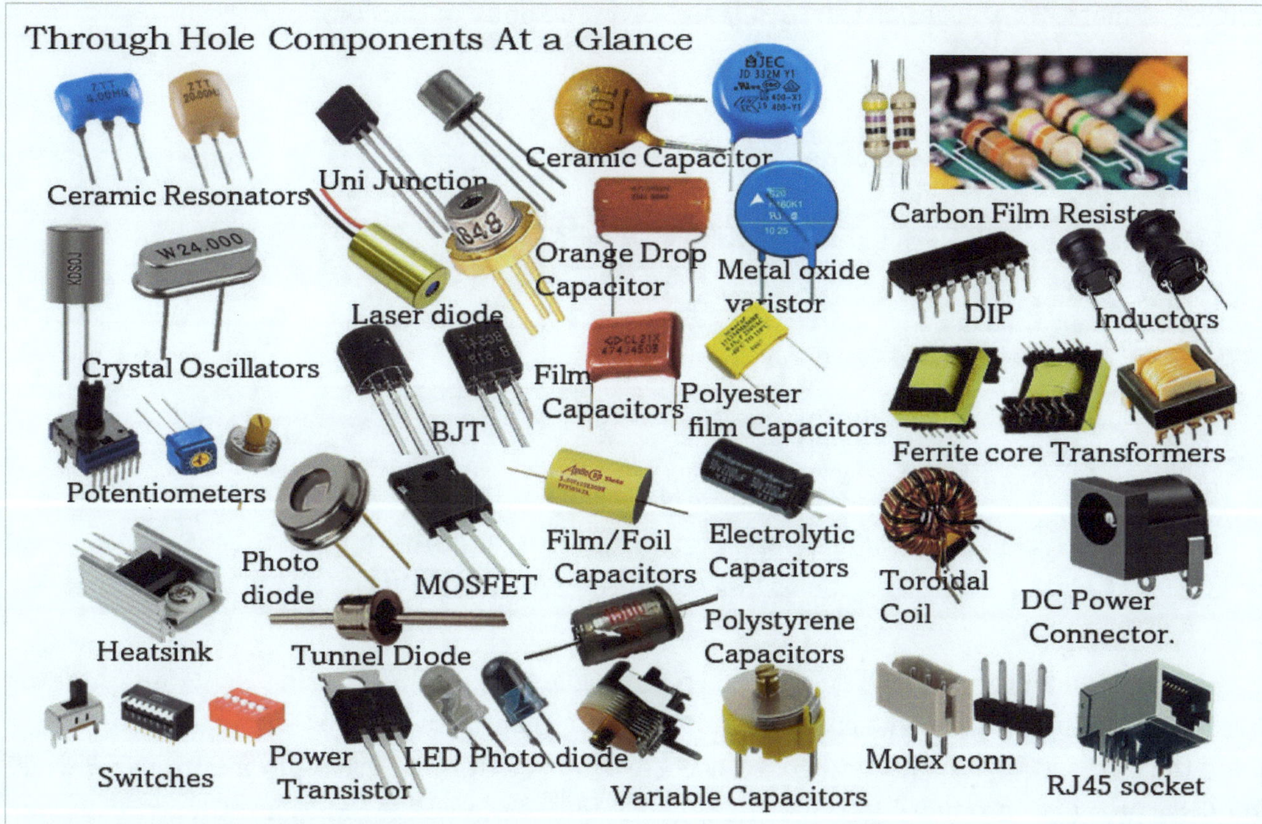

The mounting technique involves mounting components that have lead wires onto a Printed Circuit board through holes. After the insertion, component leads will be cut and clinch as per the required process, and then the leads are soldered to offer permanent mounting. Through-hole components are best used for high-reliability products that require stronger connections between layers. Through-hole components withstand mechanical, more environmental stress - extreme accelerations, collisions, or high temperatures. So, through-hole technology is still preferable in most applications and products for SMPS power supply, Navigation systems, Aerospace, defense, and industrial or special purpose applications. Due to their various lead thickness, high wattage power, and high energy storage, component dimensions are bigger when compared to Surface Mount devices. Some drawbacks like the signal must necessarily go through all PCB layers, and low density due to minimum pin diameter and only one-sided mounting.

Through-hole technology is also useful in testing and prototyping applications that sometimes require manual adjustments and replacements. Hence benefits will be easy to solder either automatically (Wave soldering) or by hand, easy to de-solder and test, and implement interconnection between upper and lower layers (vias) in on-populated hole technology.

SURFACE MOUNT DEVICES SMD

SMD components usually have very small leads or leads flush to the component body surface as if it looks like no leads at all. Leads can be either visible or non-visible upon mounting. Usually smaller than its through-hole counterparts. It may have short pins or leads of various styles, flat contacts, a matrix of solder balls, or terminations on the body of the component.

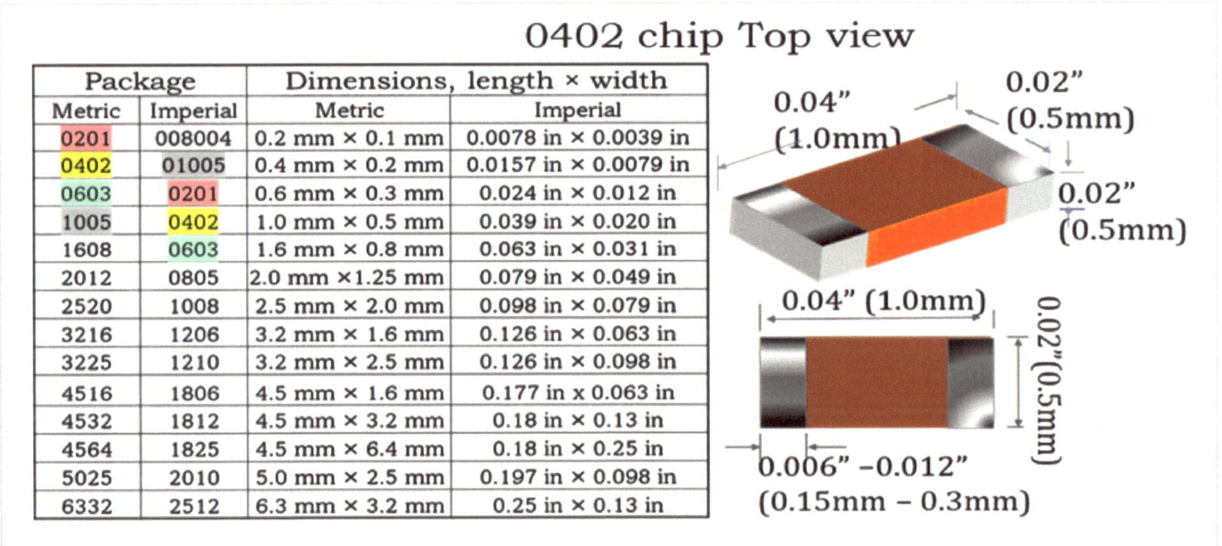

0402 chip Top view

| Package | | Dimensions, length × width | |
Metric	Imperial	Metric	Imperial
0201	008004	0.2 mm × 0.1 mm	0.0078 in × 0.0039 in
0402	01005	0.4 mm × 0.2 mm	0.0157 in × 0.0079 in
0603	0201	0.6 mm × 0.3 mm	0.024 in × 0.012 in
1005	0402	1.0 mm × 0.5 mm	0.039 in × 0.020 in
1608	0603	1.6 mm × 0.8 mm	0.063 in × 0.031 in
2012	0805	2.0 mm ×1.25 mm	0.079 in × 0.049 in
2520	1008	2.5 mm × 2.0 mm	0.098 in × 0.079 in
3216	1206	3.2 mm × 1.6 mm	0.126 in × 0.063 in
3225	1210	3.2 mm × 2.5 mm	0.126 in × 0.098 in
4516	1806	4.5 mm × 1.6 mm	0.177 in × 0.063 in
4532	1812	4.5 mm × 3.2 mm	0.18 in × 0.13 in
4564	1825	4.5 mm × 6.4 mm	0.18 in × 0.25 in
5025	2010	5.0 mm × 2.5 mm	0.197 in × 0.098 in
6332	2512	6.3 mm × 3.2 mm	0.25 in × 0.13 in

A Semiconductor package is a metal, plastic, glass, or ceramic casing containing one or more semiconductor electronic components. The package protects against impact and corrosion, holds the contact pins, or leads that are used to connect from external circuits to the device, and dissipates heat produced in the device. The majority of passive SMDs are Chip resistors or capacitors for which package sizes are reasonably well standardized, whereas active components are still not well standardized.

SMD components like Resistors, capacitors, and Inductors come in a variety of standard packages. Terminology standard packages will be used either in the Metric Code or the Imperial code. The metric code is in millimeters. Imperial code is in Inches. For example, 0402 imperial components. This means the component is 0.04 x 0.02 Inches. The same package is in metric code 1.0mm x 0.5 mm (1005). Please note that one must care while ordering 0201,0402, and 0603 packages, you must mention the package unit to avoid confusion, i.e., imperial or Metric. Because the same codes appear twice in different L x W dimensions.

Due to package size, advantages will be that – can be populated with high density, devices can be mounted on both sides of the PCB, components do not block signals in the inner layer, pins can be thinner, and a higher degree in the automation of the mounting process. Less parasitic inductance and capacitance and less Cost.

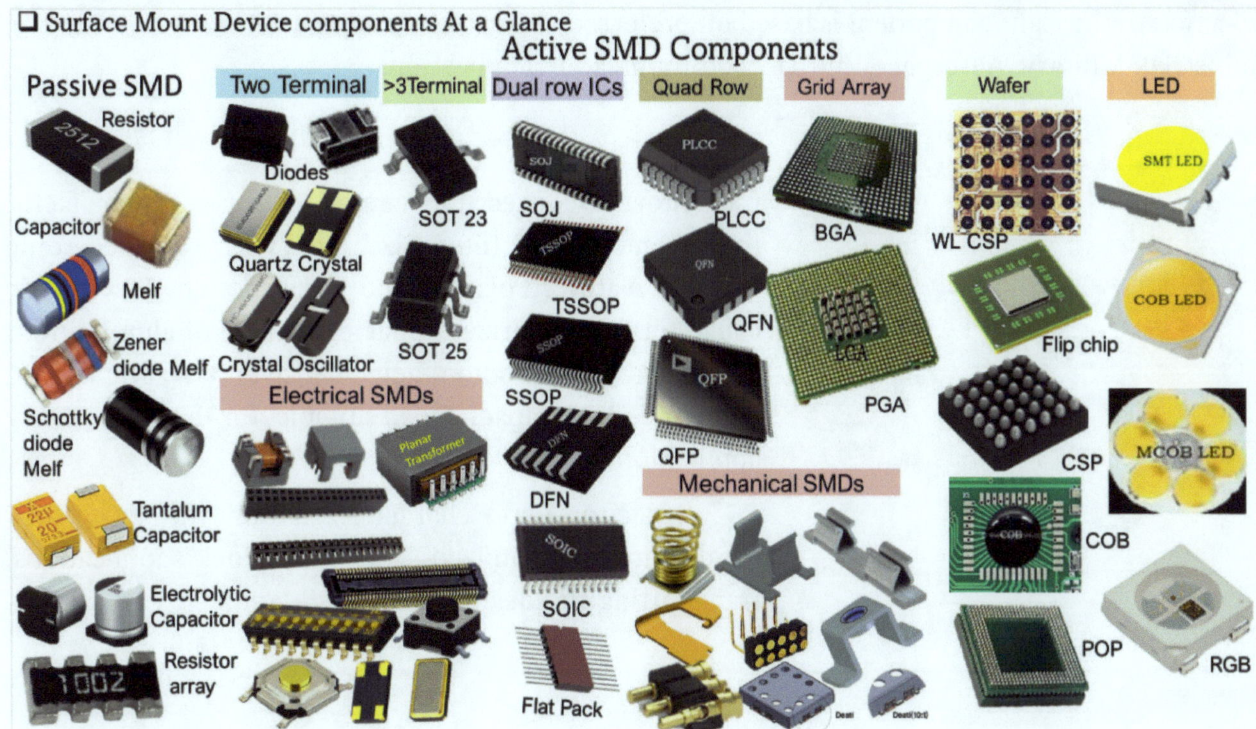

In the same way, disadvantages can be – Reliability issues due to thermal and mechanical stress during soldering, and Operation, such as different thermal expansion coefficients. Manual soldering is very difficult, and repair ability is also very difficult.

A LITTLE ABOUT PASSIVE COMPONENTS

Resistors

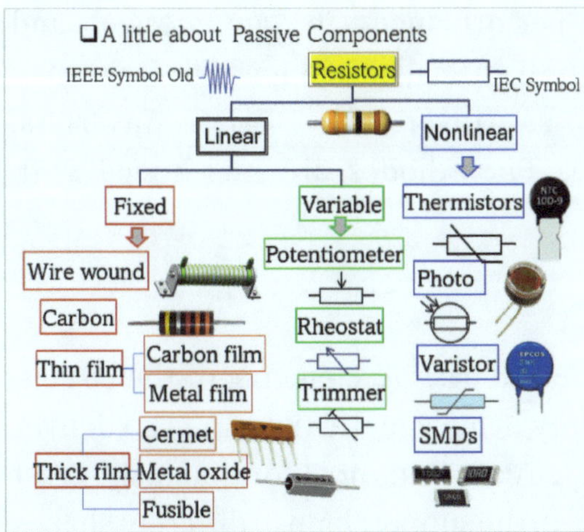

Linear - Those resistors whose values change with the applied voltage and temperature are called linear resistors. In other words, a resistor whose current value is directly proportional to the applied voltage is known as a linear resistor. Thick film, metal oxide, ceramic, Carbon film, and Trimmers are mostly used for electronic assemblies.

Non-linear resistors are those resistors where the current flowing through them does not change according to Ohm's Law but changes with a change in temperature or applied voltage. Thermistors, Varistors, Photo resistors, most common examples.

Each class of Resistors has unique capabilities such as Physical characteristics, building characteristics, and features like High power dissipation, excellent pulse load capability, precision power supply, wide resistance range, moisture resistance, Component protection, etc. Hence, the selection of the right component is based on preferred characteristics and desires for Automotive, Industrial, Military, Aerospace, and Telecommunication applications.

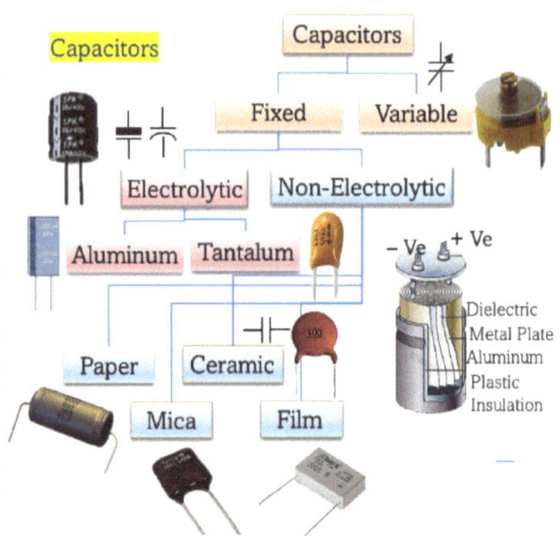

Capacitors

Capacitors are used for energy storage, power factor improvement, and filter Circuits. Types vary according to their Nature, Polarity, and construction. Electrolytic capacitors are polarized capacitors made of aluminum or Tantalum with an oxide dielectric layer. Ceramic is non-polarized, low capacitance, suitable for high-frequency applications

Capacitors are characterized for the Frequency band range, optimized for DC block applications, transmitting the desired AC signal, and offer resonance-free performance.

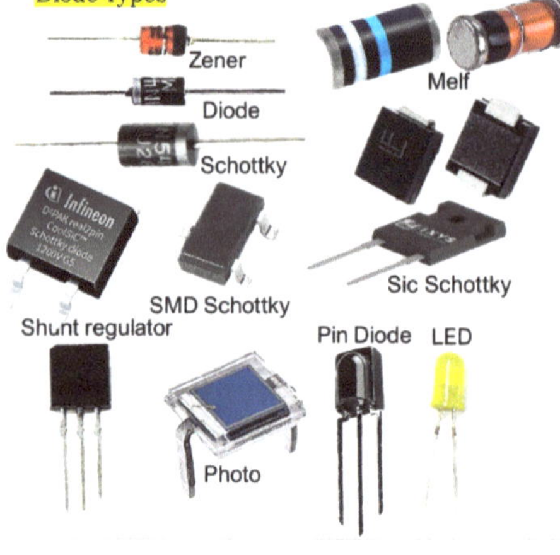

Diodes

A diode is a device that allows current flow in only one direction. i.e., from the anode to the Cathode. Various types of diodes are used in the circuit based on parameters and special applications.

For example, Zener diodes are mainly used in protection circuits or as crude voltage regulators. Likewise, Schottky diodes feature faster switching speeds. Thermal performance in automotive and industrial applications. Especially used for high-frequency applications such as Inverters, DC-DC inverters, Protection devices, and current flow regulators. whereas Photodiodes are high-speed with good photosensitivity.

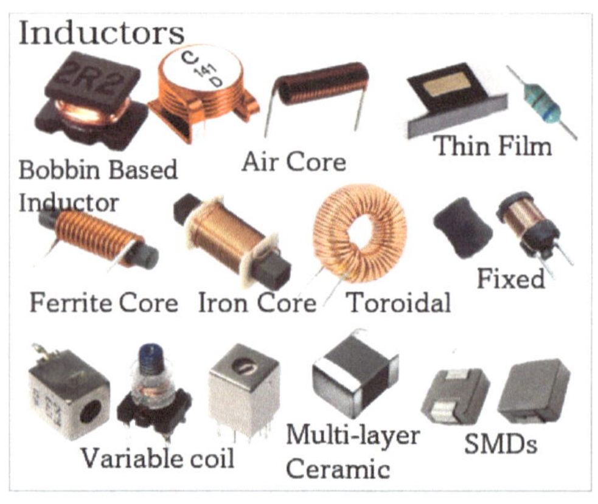

Inductors

Inductors are regarded as passive elements of the circuit, which can store energy in a magnetic field and can deliver that energy to the circuit, but not continuously. Vast varieties are available in the supply chain in different shapes and sizes for their uses. Preferably used for applications like filters for analog circuits, Motors to turn electrical energy to mechanical energy, SMPS power supplies, and lightweight transformers.

Depending on the type of material used in inductors, they can be classified as Iron Core, Air core, Iron powder, fixed, Variable, and Multi-layer Ceramic chip inductors. Apart from these, based on special applications, Bobbin-based either radial or Axial lead forms, Toroidal, and coupled molded inductors are common in electronic assemblies.

ACTIVE COMPONENT – INTEGRATED CIRCUIT PACKAGES

Three terminal packages:

Transistor

The transistor is a solid-state semiconductor device that can be used for amplification, switching, voltage stabilization, signal modulation, and many other functions. It allows a variable current from an external source to flow between two of its terminals depending on the smaller voltage or current applied to a third terminal. Transistors are made either as separate components or as part of an integrated circuit.

Two types of modern versions of transistors, BJT (Bipolar Junction transistor) and FET (Field effect transistor), are in use. BJT is a current-controlled device, whereas FET is a voltage-controlled, three-terminal unipolar semiconductor device. BJT made two types of configurations, PNP and NPN devices. Similarly, FET sorted JFETs (Junction Field Effect Transistors) are further classified into P channel, N channel depletion mode, and MOSFET Metal-Oxide-semiconductor Field Effect Transistors), which are the most used devices in the present technology. The main advantages of FET are extremely high input impedance, low power consumption, low heat dissipation, and highly efficient devices. Used in circuits such as Buffer amplifiers, measuring instruments, Voltmeters, phase shift oscillators, and especially to reduce loading effects.

This book is intended to teach you a little about Surface-Mount Technology. This section aims to learn more about electronic packages, which are most important for Hierarchy levels, and to provide more ideas about semiconductors, which are the heart of Electronics engineering. Hence, before we further learn about IC package classifications, let us understand the front-end and back-end processes of semiconductors.

SEMICONDUCTOR WAFER FABRICATION IN BRIEF

Process 1 – Wafer fabrication

As shown in the process, Silicon is extracted from Sand and cast into " INGOT " blocks. Ingots are cut into wafers at a desired thickness. Wafers were then polished and subjected to Vigorous inspection. Wafer size (Diameter) could be 8", 12", 18 inches, or even more. Wafers are fabricated in an environment 100,000 times cleaner than medical operation theaters. As a part of the Front-end process line, Chips are made on Silicon Wafers, and each chip is processed for billions of transistors and their interconnection—chips formed only on the top layer of the Wafer. As a standard practice, wafers are processed in batches of 25 called one Lot.

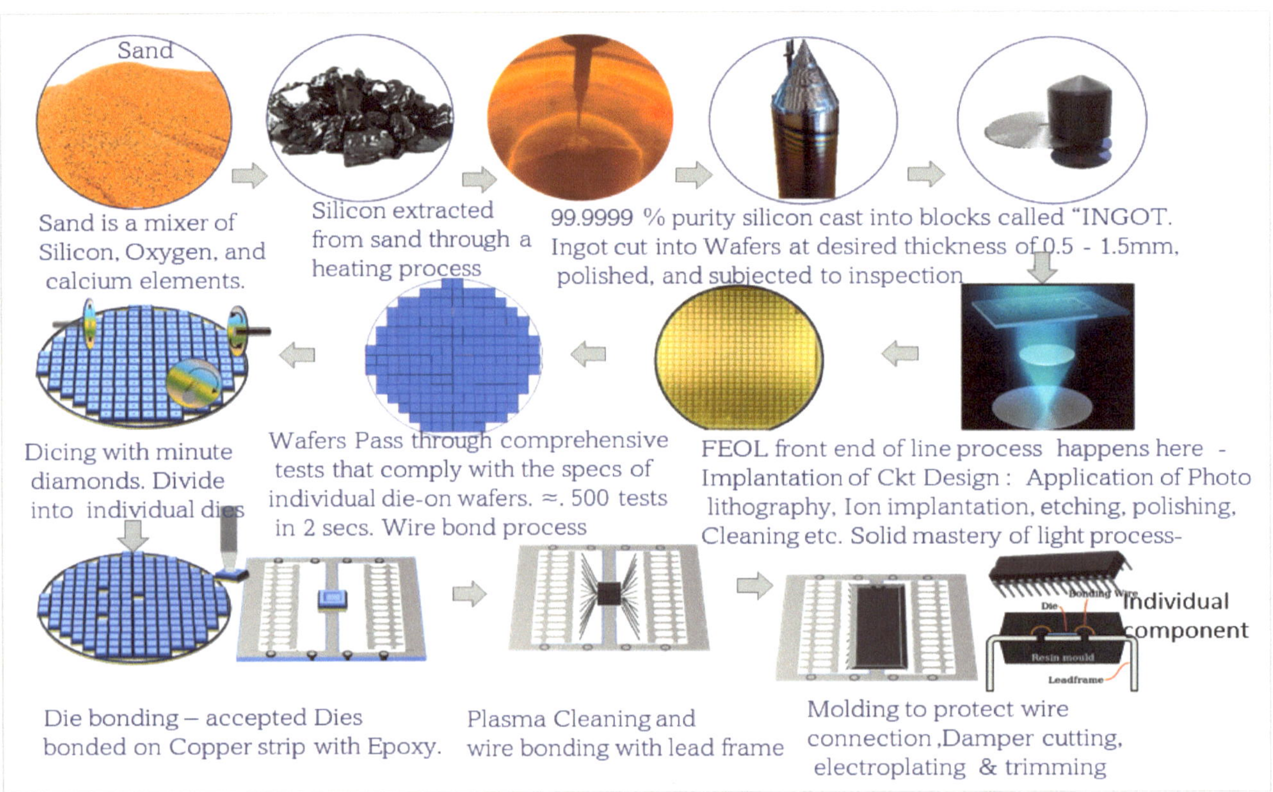

PROCESS 2 – FRONT END OF LINE

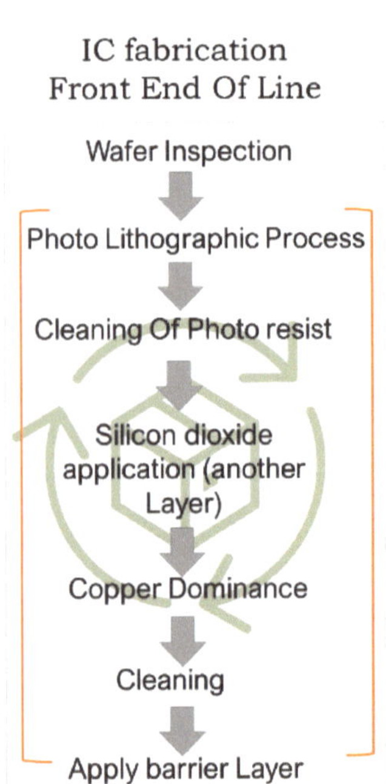

Circuit design is the first step of any IC fabrication, such as Processors, Memory chips, high graphic processors, wireless communication circuits, etc. This process involves CAD computer-aided tools to design according to the product design specifications. Once the circuit diagram is ready, develop stencil fabrication/glass mask or reticle-containing patterns that need to be transferred to silicon. This prefabricated stencil is composed of Transparent glass and Opaque chrome projected through the lens. The lens reduces the size of the stencil pattern.

The Photo Lithographic technique transfers the circuit structure to the wafer. The key to the whole process is the solid mystery of the light process. The process steps involve the application of photoresist coating (a light-sensitive polymer, thickness around 0.5μm - 1μm, soluble in organic solvent) on a wafer, transfer the circuit through UV light removal of the exposed part of photoresist by the developer, Ion implantation where electrical properties are specified.

Then doped atoms are injected into the silicon structure, distributing randomly into the silicon Lattice structure. The next process is fire. At elevated temperatures, the doped atoms become flexible and take on a fixed position in the atomic structure to make transistors.

The next process is a copper-dominant process. The finest billions of separate transistors link up with the finest interconnect conductors, and the thickness of these paths on the circuit is 500,000 times narrower than human hair.

The entire process described in the front end of the lines repeats several levels (hundreds of layers) layer by layer, several circuit structure implants happen as per circuit specifications. The cleaning process shall be mandatory at every stage of the process. Before copper is poured into the interconnects, a barrier layer is applied at every level to avoid short circuits. Hence, the front end of IC manufacturing passes through several hundred processes and takes more than a month to make a wafer ready.

PROCESS 3 – BACK END OF LINE

Silicon wafers are supplied in dust-tight boxes. All wafers pass through comprehensive tests with the aid of special test methods. Test that complies with specifications. Tests are conducted on each individual die on the wafer. Mask done for defective Die and eliminated later. ICs contain Digital and Analog circuits and require hundreds of tests (more than five hundred) in less than 3 secs.

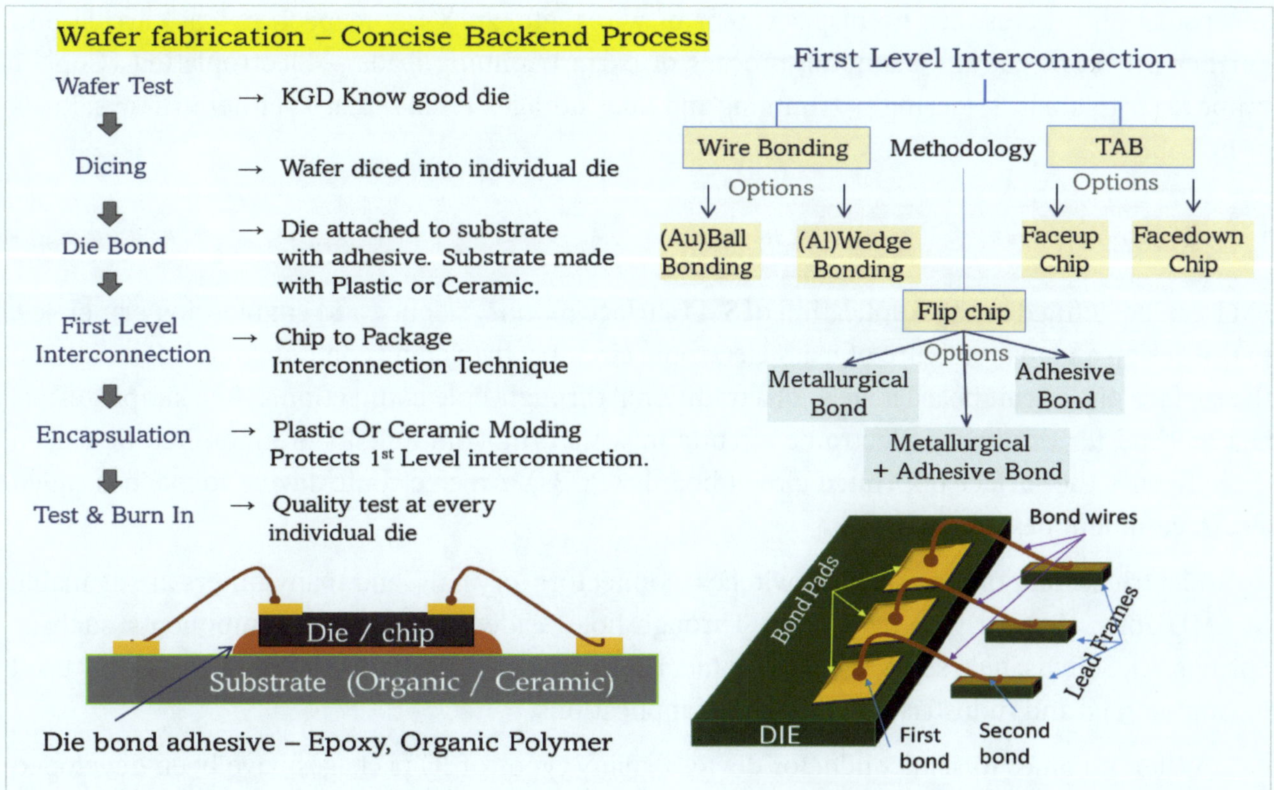

Once the inspection is over, the next process is dicing. Wafer is divided into individual dies. Separation of dies from wafers is done in different methods, the most common being the Dicing

Saw (Mechanical Saw) method or by Laser. Cutting is done with a Steel frame with minute wires fitted with minute Diamonds (only 30μm wide) rotating at greater than 30,000 rpm. Deionized water is used to cool and move the silicon dust that is created during this process.

After dicing, defective dies are segregated by an automatic machine, and good dies are bonded on copper strips with epoxy. These copper strips (known as lead frames) subsequently become electrical contacts with the printed circuit board. The design of lead frames can be either through-hole or surface mount type. In the case of multi-chips, like microcontrollers and Flash memory, two or more chips are embedded in one die. Further plasma cleaning was done to remove any organic contaminants on the lead frame, i.e., contacts.

The process of making electrical contact between the Die and the Copper Lead frame through wire bonding is called the First level Interconnection. Various wire bonding methodologies were adopted in the first level of Interconnection. Wire Bonding, TAB Tape Automated Bonding, and options for Flip Chip, most common Practices. Wire bonding is achieved through Thin gold wire (25μm thick or less), welded on bond pads through a fully automated machine at a temperature of 200°Cwith the aid of pressure and ultrasonic energy. In the case of multichip, hundreds of connections are required in seconds. Once interconnections are completed, extensive quality tests are carried out, e.g., mechanical pulling test.

To protect the sensitive wire connections (First level Interconnections) from mechanical damage and to prevent the silicon die from corrosion, all components are molded with plastic or ceramic compound at high pressure at approximately 175°C. To ensure no changes in this encapsulation process, i.e., no displacement of wires, etc, an X-ray inspection was carried out. Further process - cut and stamping process of extra trimming leads – Electroplating (Copper frame leads to dip in Soldering) – trimming and final storage of individual IC either in Trays, Reels or in Tubes.

ELECTRONIC PACKAGE CLASSIFICATIONS – SMD

SMT can be defined as the Application of SM (surface mount) science and engineering principles, to the design and manufacture of mass electronic circuitry by placing components and devices on the surface of a circuit board instead of a traditional through-hole connection. OR in simple terms, is a method for producing electronic circuits in which the components are mounted or placed directly onto the surface of printed circuit boards (PCBs). An electronic device so made is called a surface-mount device (SMD).

Electrical components such as switches, connectors, crystals, and many others are available in SMD form. Similarly, conventional Through-hole leaded Mechanical components such as Springs, RF Shields, heat sinks, Battery contacts, etc., are converted into SMD and are widely used in Commercial and Industrial PCB assembly applications.

When it comes to semiconductor devices, many types of IC packages have been developed since the year 1964, starting from 14-pin DIP Dual in package to the present era of System in package (stacking of ICs) and further developments. All these developments worked out to overcome reliability issues (usage of best packaging material, plastic, glass, ceramic casing), Space utilization (reduce PCB size), and reduce the height of the package so that product size can be reduced in terms of width, height, and to reduce footprint on PCB.

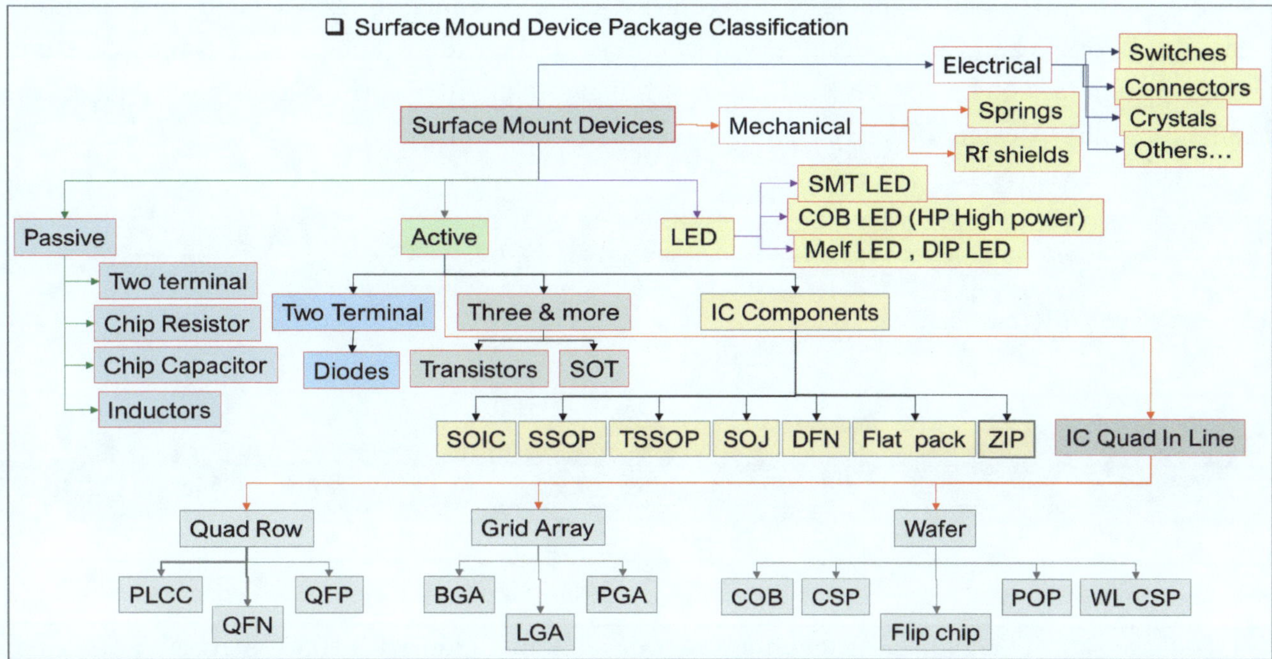

IC Packages are classified in terms of 1. Mounting type – SMD, through a hole, and contact mounting type (tape carrier package) 2. Terminal direction- i.e., leads on one side, dual, or Quad, Terminal 3. shape – i.e., Lead shape (L shape for through hole, Gullwing, J type, electrode bud (no lead), solder ball (bump, a terminal pin type) and 4. other classifications abbreviated, given formal names based thermal abilities, body material type, body dimensions such as Body length, width, thickness, lead length, lead width, lead pitch, bump diameter and bump pitch. Information of all these classifications of a package is very important, and all this data should be required or fed to the relevant production types of equipment in PCB assembly manufacturing.

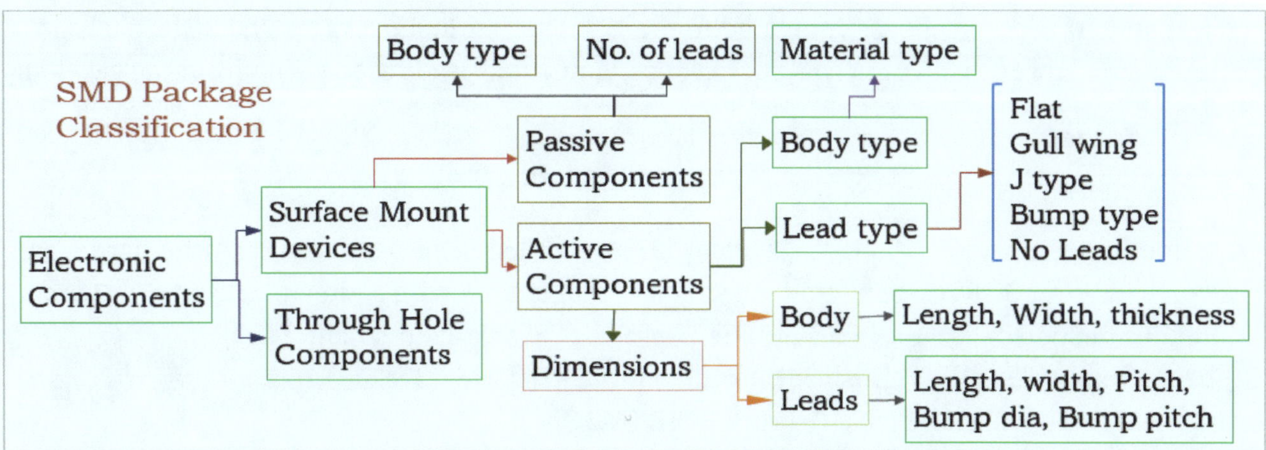

When it comes to internal architecture, many Integrated Circuits are in the market, for example, Power management ICs, Analog ICs, Digital ICs, Mixed Analog and Digital ICs, Memory Chips, Interface ICs, Programmable devices, Microprocessors, microcontrollers, and much more. Chip makers are many globally. Many top semiconductor foundries manufacture ICs on behalf of Clients in the fields of Automobile, Consumer electronics, Smartphones, medical equipment, etc., and general-purpose electronics. In most cases, the same internal Architecture ICs are manufactured and available in different packages as shown in the Package classification. Product developers and Circuit designers must choose and decide which package is most suitable to their product, keeping the given Class of product, Reliability, optimization cost, design for manufacturability, design for serviceability, and outsourcing capability.

FOUR OR MORE TERMINAL PACKAGES

SOT –Small Outline Transistor

SOT package is a rectangular, surface mount transistor or diode with three or more gull wing leads, a plastic molded package. The leads are on two lengths of the package. Popular sizes are the SOT23, SOT143, SOT223, and SOT89. Device power dissipation is around 0.6 watts. Abbreviation and nomenclature numbers can be based on SOT + Lead Pitch + (lead span x lead height) – Pin quantity. This type of package is commonly used in consumer electronics products. SOT23 has 5,6,8 pin versions. Unlike other SOT packages, SOT23 does not have a thermal pad. Therefore, to achieve good thermal dissipation, PCB design should be optimized.

SOIC – A Small Outline Integrated Circuit

SOIC is a Surface–mounted integrated Circuit (IC) package. It occupies an area of about 30–50% less and 70% less thickness compared to the DIP Dual-in-line package. The SOIC package is shorter and narrower than DIPs. This package has "gull wing" leads protruding from the two long sides and a lead spacing.

SOP – Small outline package

SOP is another name for SOIC. JEDEC standard USA is called SOIC whereas JEITA standard Japan is called SOP. After SOIC came a family of smaller form factors with a pitch of less than 1.27 mm. SOP family includes - Plastic small outline package (PSOP), Thin small outline package (TSOP), Thin-shrink small outline package (TSSOP), shrink small outline package (SSOP), Mini-SOIC (MSOIC).

SSOP – Shrink small outline package

Chips have "gull wing" leads protruding from the two long sides and a lead spacing of 0.0256 inches (0.65mm). 0.5mm lead spacing is less common but not rare. Semiconductor families such as operational amplifiers, drivers, optoelectronics, controllers, logic, analog, memory, comparators, and more using Bic MOS, CMOS, or other silicon / GaAs technologies are well addressed by the SSOP product family.

MSOIC – mini SOIC

Another SOIC variant, available only for 8-pin and 10-pin ICs, is the mini-SOIC, also called micro-SOIC. This case is much smaller, with a pitch of only 0.5mm

SOJ – Small outline J Leaded Package

Developed to reduce space required on substrate to replace DIP Dual in Package. Leads wind towards the internal side like a "J" shape so that tips hold the package body. SOJ is a version of SOIC with J-type leads instead of gull-wing leads.

TSSOP – Thin Small Outline Package

TSSOP is a rectangular, thin-bodied component. The ICs on DRAM memory modules were usually TSOPs until they were replaced by a ball grid array (BGA). TSSOP's leg count can range from 8 to 64.

PLCC – Plastic Leaded Chip Carrier

Released in 1976, but did not seem to market adoption. The PLCC utilizes a "J"-lead with a pin spacing of 0.05" (1.27 mm). PLCC packages can be square or rectangular. A PLCC circuit may either be installed in a PLCC socket or surface-mounted. A PLCC socket may be necessary in situations where the device requires stand-alone programming. It is particularly common for read-only memories as it provides an easily swappable socketed chip. Applications range from consumer products to automotive and aerospace.

DFN/QFN – Dual /Quad Flat No lead package

It is the most popular semiconductor package. Low cost, small, good electrical, and good thermal performance. Pins on four edges of the bottom surface of the Package. Introduced to replace the gullwing lead. Component leads are embedded in the plastic, i.e., the electrode pad is prepared as the terminal for the connection so that pins cannot be bent during handling. QFNs have an exposed thermal pad on the bottom of the package that can be soldered directly to the system PCB for optimal thermal transfer of heat from the die.

QFP - Quad Flat Package

QFP is a surface mount integrated package with a "gull-wing" that allows solid footing during assembly to the PCB. Versions ranging from 32 to 304 pins with a pitch ranging from 0.3 to 1.0 mm are common. Package thickness ranges from 1.4mm to 3.6mm. Dimensions vary from 4mm X

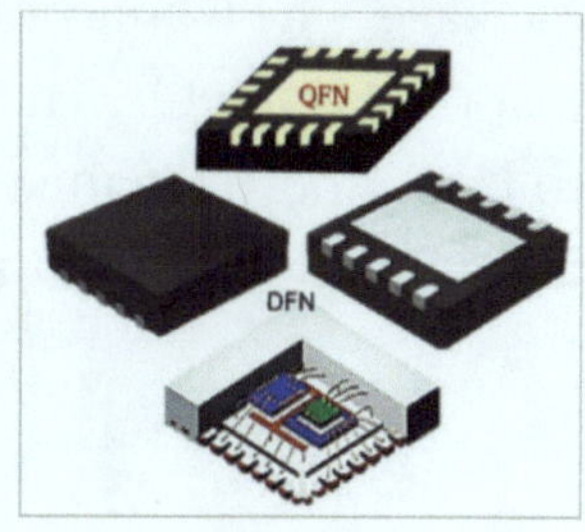

4mm to 32 mm X 32 mm. It is commonly used for NOR Flash memories and other programmable components. The following types also come in the QFP family.

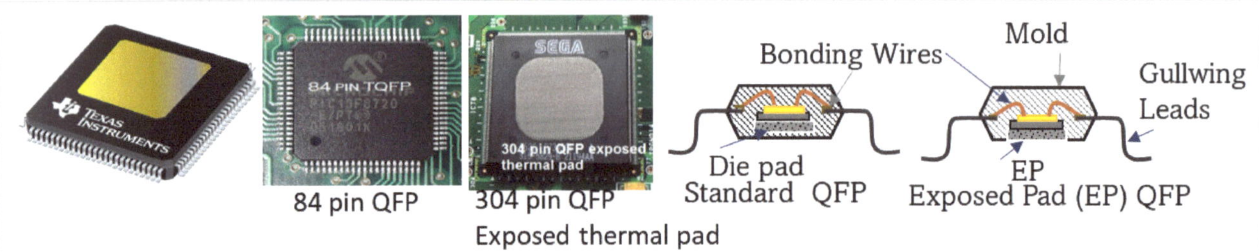

PCB DESIGN GUIDELINES AND REQUIREMENTS FOR QFP

PCB Pad design was designed according to IPC standard IPC-7351. A proper PCB footprint and stencil designs are critical to surface mount assembly yields and subsequent electrical and mechanical performance of the Mounted Package. Recommended Pad finish: HASL, OSP, ENIG

Heat transfer / Electrical grounding from QFP package (exposure pad EP) to the Board through solder joint. Thermal vias are necessary to effectively conduct from the surface of the PCB to the ground Plane (s). Stencil thickness depends on the lead pitch, package Coplanarity, etc. A standard thickness of 0.1mm can be varied from 0.13mm to 0.2mm, depending on the pitch. Nitrogen reflows are recommended to improve solder ability and

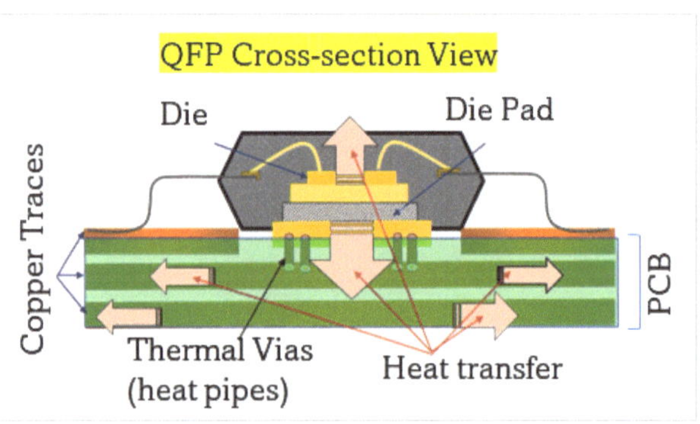

reduce defects. IPC/JEDEC –J STD- 033 is being followed for floor life, storage conditions, and handling precautions after the original package is opened. When more than one QFP supplier or Source is expected, PCB Layout should be optimized for both Parts.

These are Electrostatic Discharge Sensitive devices. Proper precautions are required for handling and processing. These parts are supplied in Trays, carrier tape, or reels as per JEDEC, EIA – 4813 design standards. The recommended Backing temperature of the tray is 125°C. Components must be mounted and reflowed within the allowable time (Life out of Bag).

BGA - Ball Grid Array

A ball grid array (BGA) is a type of SMD device packaging used for Integrated circuits. BGA packages are used for permanently mounting devices such as Microprocessors and microcontrollers. A BGA can provide more interconnection pins than a QFP and can be put on a dual in-line or a Flat package. The whole bottom surface of the device can be used for interfacing contact pins instead of just the perimeter. The leads are also, on average, shorter than with a perimeter-only type, leading to better performance at high speeds. The soldering of BGA devices requires precise control and is usually done by automated processes. BGA devices are not suitable for socket mounting.

Ball Count	Total	4	≈	2500	Ball Pitch	Fine	0.5mm	0.75mm	0.8mm
	Matrix	2× 2	≈	50 × 50		Std	1.0mm	1.27mm	1.5mm

Ball Diameter	Pitch	1.27mm	1.5mm	1.0mm	0.8mm	0.5-0.75
	Dia.	0.75mm		0.6mm	0.5mm	0.4mm

BGA Types	µBGA	CSP	PBGA	CBGA	CLGA	LBGA	S BGA
	micro	Chip scale	Plastic	Ceramic	Ceramic Land	Laminate	Thermally enhanced

Advantages of BGA

High density, Heat conduction, and low inductance leads. With discrete leads, lower thermal resistance between the Package and PCB allows heat generated by the integrated circuit inside the package to flow more easily to the PCB, preventing the chip from overheating. In the same way, Low inductance leads to lower unwanted inductance, a property that causes unwanted distortion of signals in high-speed electronic circuits. BGAs, with their very short distance between the package and the PCB, have low lead inductances, giving them superior electrical performance to pinned devices. The BGA can be refurbished (or reballed) and re-installed on the circuit board.

Disadvantages of BGA

They are not mechanically compliant. Mechanical stress issues. Additional process "Under filling Mandatory. Any difference in the coefficient of thermal expansion between PCB substrate and BGA (thermal stress) or flexing and vibration (mechanical stress) can cause the solder joints to fracture. RoHS lead-free solder alloy assemblies have presented some further challenges to BGAs, like the "head in pillow" soldering phenomenon. It is difficult to find soldering faults. X-ray machines are required to look underneath the soldered package. The BGA Rework station must be for BGA Repairs (Removal and Placing).

POP - Package on package

POP is an integrated circuit packaging method to combine vertically discrete logic and memory ball grid array packages. Two or more packages are installed atop each other, i.e., stacked, with a standard interface to route signals between them. This allows higher component density in devices, such as cell phones, personal digital assistants (PDA), and digital cameras. Widely used in applications for memory stacking, i.e., two or more memory packages are stacked on each other. Mixed logic-memory stacking, i.e., logic (CPU) package on the bottom, memory package on top. For example, the bottom could be a System on a chip (SOC)for a mobile phone. The logic package is on the bottom because it needs many more BGA connections to the motherboard.

CSP - A chip-scale package

chip scale, the package has an area no greater than 1.2 times that of the die. Must be a single-die, direct surface mountable package. These packages as CSPs are their ball pitch should be no more than 1 mm. Flip chip bond at 240°C. Underfilling cure at 150°C. Lid attachment at 150°C and Ball mount at 220°C

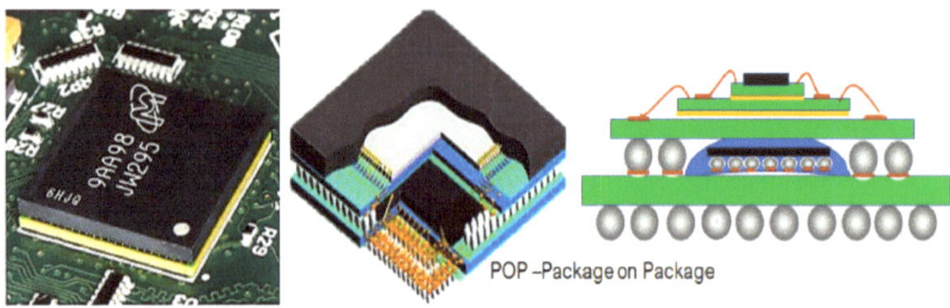

WL –CSP   Wafer level chip scale Package

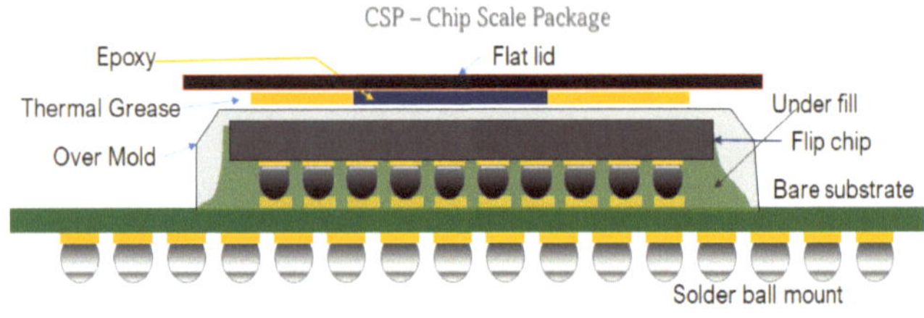

The die may be mounted on an interposer upon which pads, or balls are formed, like with flip chip ball grid array (BGA) packaging, or the pads may be etched or printed directly onto the silicon wafer, resulting in a package very close to the size of the silicon die: such a package is called a wafer level package (WLP) or a wafer-level chip-scale package (WL-CSP).

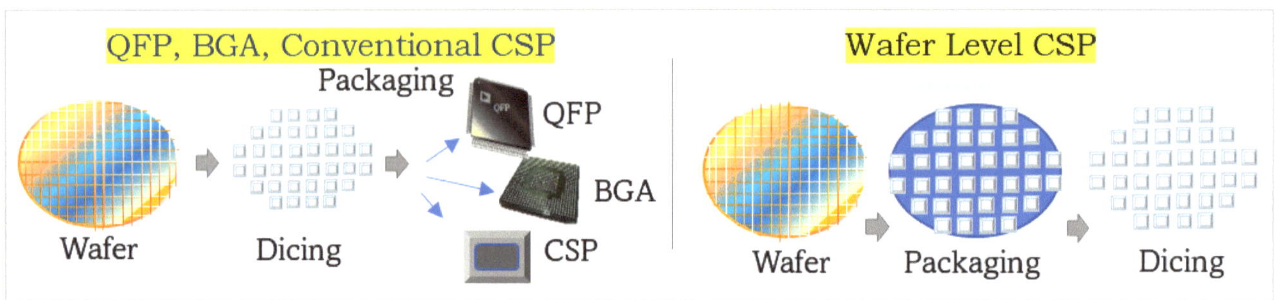

Wafer-level packaging consists of extending the wafer fab processes to include device interconnection and device protection processes. Most other kinds of packaging do Wafer, dicing first, and then put the individual die in a plastic package and attach the solder bumps.

A major application area of WLP is smartphones due to the size constraints. For example, the Apple iPhone 5 has at least eleven different WLPs, the Samsung Galaxy has more than six WLPs and the HTC 1X has seven WLPs. Functions provided by WLPs in smartphones include compass, sensors, power management, wireless, etc. It has recently been rumored that the new models of the iPhone will use fan-out wafer-level packaging technology to achieve a thinner and lighter model.

Flip Chip

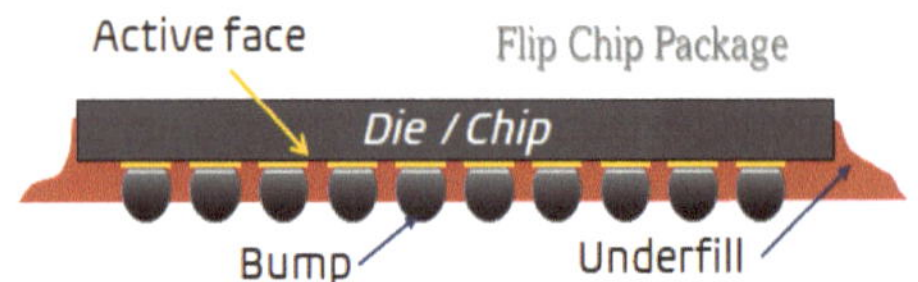

Also known as Controlled Collapse Chip Connection - a method for interconnecting semiconductor devices such as IC chips and microelectromechanical systems (MEMS), to external circuitry with solder bumps that have been deposited onto the chip pads.

The solder bumps are deposited on the chip pads on the top side of the wafer during the final wafer processing step. To mount the chip to external circuitry (e.g., a circuit board or another chip or wafer), it is flipped over so that its top side faces down and aligned so that its pads align with matching pads on the external circuit, then the solder is reflowed to complete the interconnect. This contrasts with wire bonding, in which the chip is mounted upright, and wires are used to interconnect the chip pads to external circuitry.

SUMMARY

The dual in package was first introduced in 1964 by Fairchild. It is a 14-pin through-hole package. Further developments introduced 4-pin to 64-pin DIPs with plastic or ceramic mold, especially used as memory modules. Unlike fabless foundries (they only design), IDMs, Integrated Device Manufacturers, e.g. Intel, IBM, Texas Instruments, etc., came up with all 4 verticals: Design, Manufacturing, Packaging, and selling as a business objective.

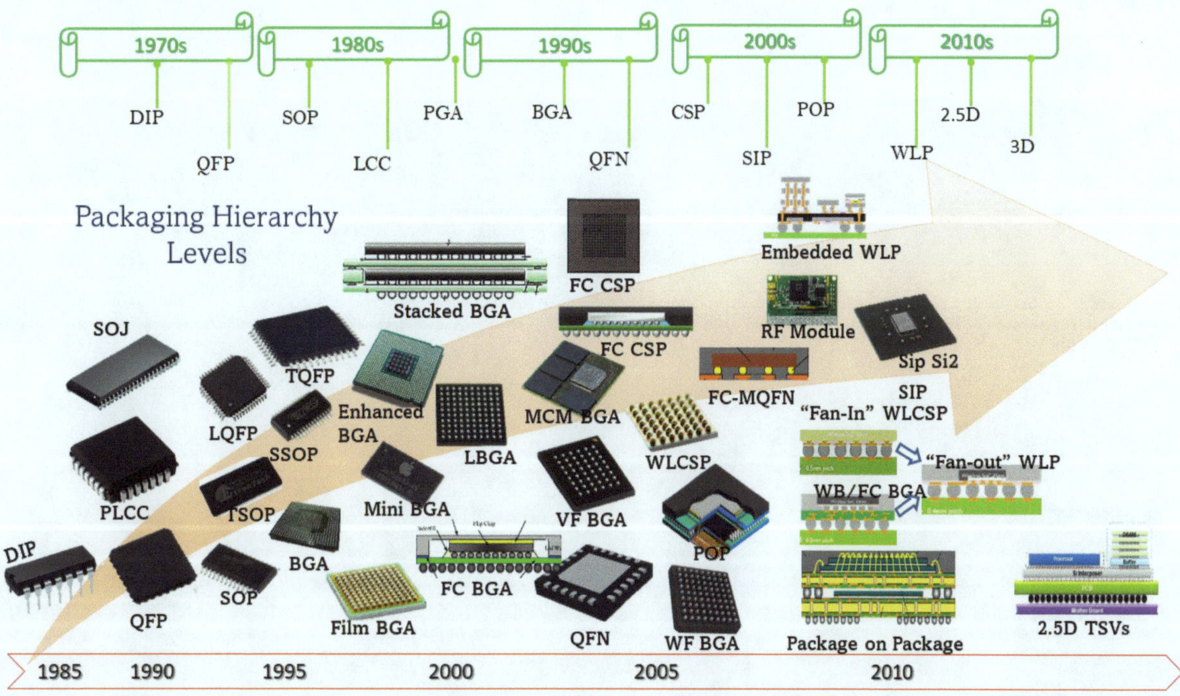

Single-die packaging was introduced in the 1970s with single functions and Systems on chips with more functions in a single die, such as processors, memories, Input /Outputs, etc. Multi-chip modules were developed in 1980. Another format is multifunctional, higher yield than a system on chip with more functionality, and with a simplified design. Different chips are integrated on a single substrate to create more functionality. From 2010 onwards, the System in package Sip (stacking of ICs) was introduced, where ICs were stacked horizontally and vertically integrated. Started the wire bond technology implies 2.5D and 3D technologies with a higher density of Transistors and contains active and passive components. Further developments of the SOP System on Package, which contains passive components embedded into a thin film substrate.

With the advent of change in Extreme Ultraviolet Lithography technology, 7nm (seven Nanometer) node test chips enable up to 20 billion transistors on a chip the size of a human fingernail. According to the IBM Press, cloud computing, big data systems, Cognitive computing, and mobile products are targets of this technology. Potential for System on-Chip applications, in addition to the CPU, you also have memory and other Logic elements on the same chip. Possible to D Ram on this 7nm chip.

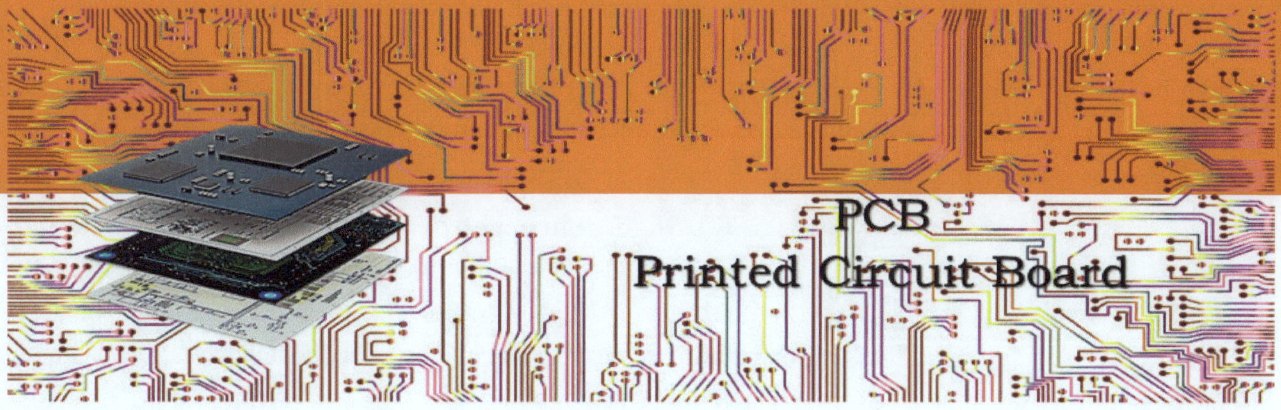

The process of circuit design can be anything. It can either start with a small Toy or the Space Station. An application can be anything like Entertainment, Defense, Communications, Space, Medical, Automotive, Automation, etc., where Logic, Memory, Controls, and decision-making are applied. Designers identify what input and what output are required. Solutions for problems and solutions for specific requirements or demands identify the basic requirements of circuit design. Complex Circuit Design starts with the thought process of Investment, investment return, Design life, and challenges in product development.

Designing a complex circuit is one aspect of the challenge, whereas transforming Circuit Design into PCB design is another aspect of the real challenge. i.e., circuit design transforms to physical forms in an electric circuit will take place. A perfectly developed PCB design is always key to the success of the manufacturing process, and a bad PCB design leads to wastage, rework, and product failure, or it may end up with a PCB design that does not translate well into the real world.

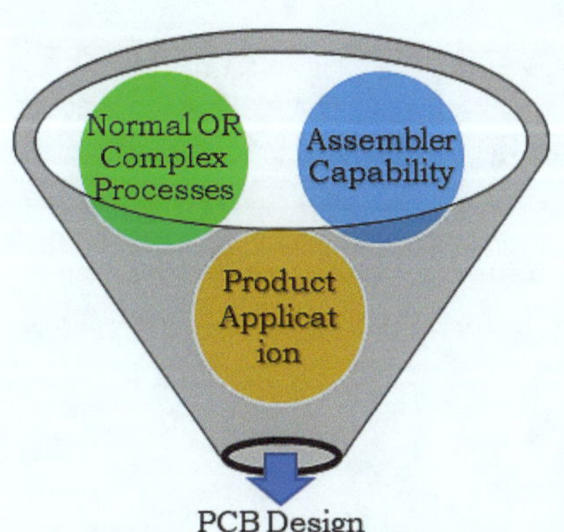

PCB designers must have good knowledge of the manufacturing process and design for Manufacturing. Knowledge to position hundreds of components and thousands of traces into a design that meets the entire range of physical and electrical requirements.

PCB designer considers the concepts of Design for Assembly (DFA) and follow the standards while designing the PCB, which refers to the Cost and efficiency for that product. Standardization offers minimum risk, clarity, and simplification. Even Assembler capability is also important for the Overall picture of DFA.

Substrate material FR4 is quite commonly used for normal and complex PCB designs, and for high-speed RF capability designs, the Polyamide material is used as a substrate. In determining the choice of substrate material, the designer must have a strong understanding of the environmental conditions that his PCBA assembly should withstand

Complex PCB designs differ from Normal PCB designs because some additional complex processes are applicable for that product in terms of via in Pad (required for fine pitch parts µBGA), wire bonding, wave soldering, onboard IC programming, Conformal coating, surface finish, etc. It might have some special Assembly requirements such as Mechanical component assembly, press-fit parts, adhesives, wire harness, enclosure assemblies, wire harnessing, and functional test points. Hence, design considerations must consider all these. Accordingly, Electronic component placement details will be organized.

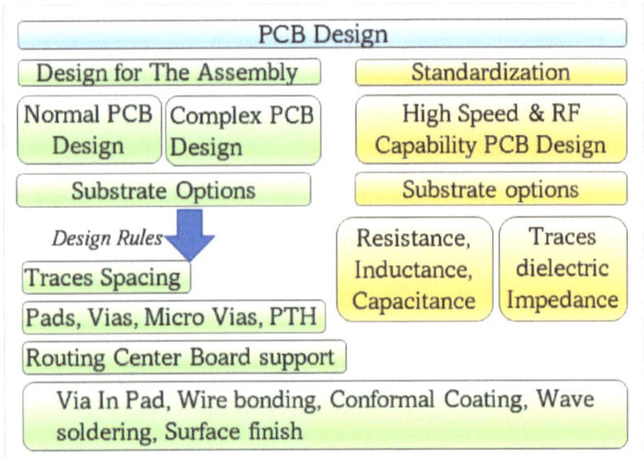

Design concepts are the same for all PCB design processes, but design rules differ for high-speed RF capability PCB design, as these designs must consider Resistance, Inductance, and capacitance. Also, traces and dielectrics affect signal rise times and impedance, limiting the upper frequency.

Similarly, PCB design rules for the wave soldering process differ in designing component orientation and pad shapes. Because pad shapes are the main concern of fine pitch surface mount parts, and shadowing concern for part orientation. The reason for double-sided PCBs is that the designer prefers to allocate placement positions for complex components such as BGA, CSP, QFP, DFN, and POP on the Top side of PCB to avoid direct contact with the wave. If the process is a must, then the package should be rotated 45° relative to the direction of travel of the wave.

Standards for PCB design, IPC -2221, Specifications for performance and accessibility, IPC – 6011, and IPC 6012 are popular standards that provide guidelines for PCB design. Many PCB design software applications (PCB Layout SW) are available, such as Windows-based packages CAD SOFT, Eagle PCB trace, etc.

INTRODUCTION

The printed circuit board is the most common name but may also be called "printed wiring boards" or "printed wiring cards". The first PCB patent was developed in 1903, and the application of single and double-sided boards continued as non-plated through holes till 1946. Further, in 1947, double-sided plated through holes developed, and from 1960 onwards the multilayer process developed.

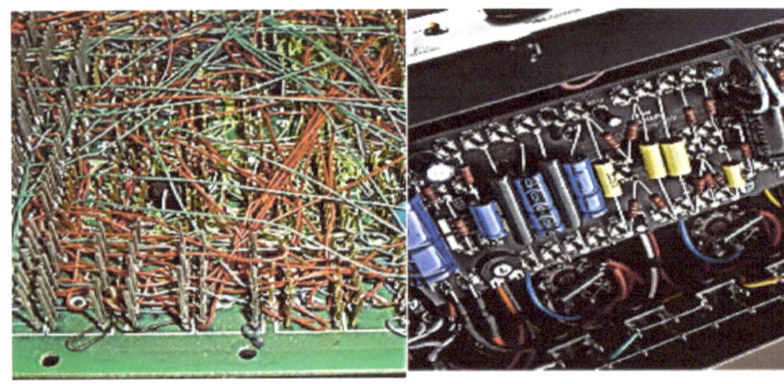

Before the advent of the PCB, Circuits were constructed through a laborious process of point-to-point wiring on a base. This led to frequent failures at wire junctions and short circuits when wire insulation began to age and crack. A significant advance was the development of wire wrapping, where a small gauge wire was wrapped around a post at each connection point, creating a gas-tight connection that is highly durable and easily changeable. This wire wrapping was only one option before the PCB was born.

PRINTED CIRCUIT BOARD

It is a base board for physically supporting and wiring the Electronic, electrical, and mechanical components. It provides mechanical support for components placed on the Board base and electrically connects them through conductive traces and pads. It is made up of selected substrates with layered copper tracks that act as wire for which components are connected. PCB is soldered to create a stronger hold for components. It consists of insulating layers and copper layers, which contain the signal traces and the powers and grounds.

Traditionally, components were mounted on the top layer in holes that extended through all layers. These are referred to as Through Hole Components (THC). With the near-universal adoption of Surface Mount Devices (SMD), you commonly find these components mounted on both the top and the bottom layers of the Printed Circuit Board.

PCB CLASSIFICATION

Printed Circuit Boards can be classified by considering the factors of fabrication processes as well as substrate materials. PCBs can be referred to as Layer-based, i.e., the number of layers, the type of substrate material used to fabricate, i.e., organic or Inorganic, how they look physical or designed, i.e., Rigid, Flexible, or semi, and the conductor pattern designed based on application.

Attributes like inorganic base, discrete wiring, additive, multi-wire, and wire wrap were phased out as the technology

PCB Classification

Layer Based	Substrate Based	Application Based	Technology Based
Single Sided PCB	Rigid PCB	Low Frequency PCB	Surface finish
Double Sided PCB	Flexi Rigid PCB	High Frequency PCB	Aluminum PCB
Multilayer PCB	Flexible PCB		Metal Core PCB
			Carbon Film PCB

was superseded by advanced technology in all aspects. Before getting into the details of PCB classifications, let us understand PCB terminology.

PCB Structure and Terminology

Transforming application design into printed circuit boards is one complex technology, and designing the PCB while maintaining the standards of Via hole, pad, anti–pads, lands, and traces is another complex technology. PCB structure should match the designer's recommendations, such as the Fabrication of PCB size, track width specifications, minimum clearance between tracks, etc.

Job orders provide guidelines about pad dimension requirements, via requirements, fill requirements, and Maximum primitive dimensions so that a designer might take these into various technical considerations. For example, if a track of 150 mils is required, a 100-mil track and a 50-mil track can be placed side by side with a 5-mil overlap to get a 150-mil track.

Traces & Spacing between Traces

A trace is a piece of copper that makes an electrical connection between two or more points on a PCB. Traces carry current. The process of laying down traces on PCB can be either plated or etched away on the surface of the substrate to leave the desired pattern. The etching method is most common in the PCB manufacturing industry.

The thickness of the Copper on PCB is specified in ounces (Oz) / Square foot. In general, ½ Oz and 1 Oz copper is the most common. Thicker copper, up to 6 Oz, is used for high current and reliability designs. The design of trace width, trace thickness, spacing between traces, layout of signal traces, power traces, and ground traces, and routing within the PCB layout, are all based on design calculations and design rule standards.

WHAT ARE SUCH CALCULATIONS?

Track /trace width design based on individual current carry and maximum temperature resistance. The resistance generates heat as the current is conducted. The trace will dissipate this heat based on surface area, airflow, and solder mask thickness. Wider traces produce less heat so that such heat dissipates easily.

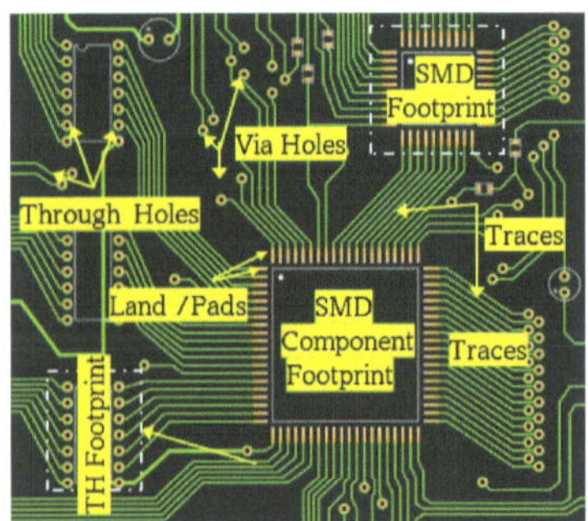

Spacing between traces, in other words, electrical clearances, is most important, and it recommends as much space as possible. The spacing between tracks /traces and upper trace width will be determined by the current flowing through it combined with the maximum temperature rise of the trace.

IPC has a set of standards about clearances and spacing charts based on environmental conditions (not voltage). Normally, Signal traces maintain minimum width, whereas power and ground trace widths are bigger, and spacing can be 60μm -40μm. Part-to-edge spacing is important for the depanelizing process; a minimum spacing of 125 mils is recommended.

Pads or Anti-Pads

Pads are small areas of copper used to connect a component pin. Anti-pad means the predetermined shapes are removed from the copper. Generally, an anti-pad is used around a via to isolate it from a power plane without connection.

Lands

Pads are needed to solder a component to the top or bottom layer of the printed circuit board, it may be referred to as a land. Pad size, shape, and dimensions depend on the Type of component package being used and the Manufacturing process being used to assemble the board. SMT Pad to lead aspect ratio, THT hole to lead aspect ratio, THT annular ring, package sizes, and Pitch defined by the IPC standards and design rules.

Sometimes, the manufacturer defines his choice. For example, the BGA pad size depends upon the surface finish. For the HASL surface finish, the BGA pad size will be 12 mils ("mil" = 0.001 inch) in diameter, and for other surface finishes, a minimum 10 mil diameter will be maintained.

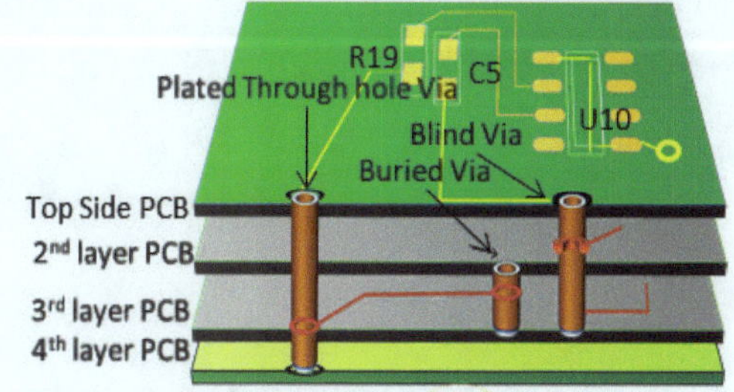

Via Holes

A Via hole is a piece of metal that makes the connection between layers on the PCB. Connects the traces from one side of the board to another by way of a hole in the board. PTH (Plated through hole), Micro vias, and via in pad all come under the category of vias.

It carries signals or power between layers through PTH. The size of the Via depends on the trace width. There are some micro vias used in between layers called Buried Vias and between the outer layer and inner layer called Blind Vias used when high-density inner connections are required.

Single-sided Layout PCB

Single-sided PCBs contain only one layer of conductive material and are best suited for low-density designs. Parts are laid out on one side, and the circuit is on the other side. It is restricted in the circuit design because there is only one side conductor, and no cross is permitted; each line must have its path.

Although the single-sided PCB is the most basic and the starting point of printed circuit technology, and indeed, the starting point of the invention, it still plays a major role in the industry. Low cost, especially for volume production. Suitable for simple circuits. Single-layer PCBs have a relatively wide field of applications ranging from power supplies, relays, sensors, and LEDs to

calculators, printers, coffee makers, and electronic toys. However, single-sided PCBs feature some performance limitations.

Double-sided PCBs

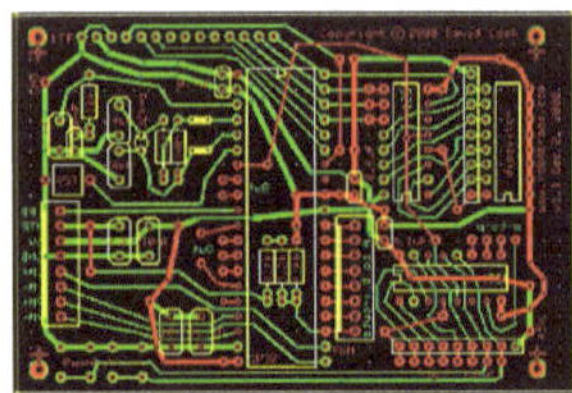 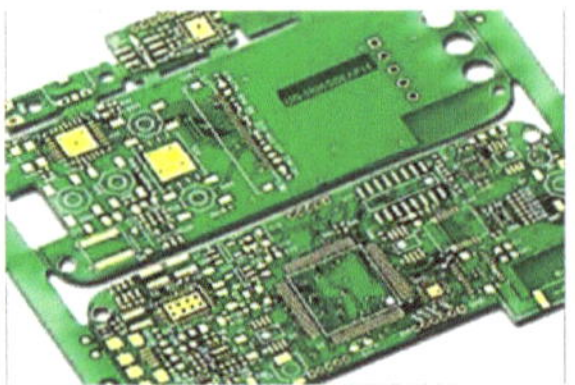

The most common and the most widely used type of board. Double-layer PCBs have two conductive layers that are placed on both sides of the substrate and so are components. Parts and components are attached to both sides of the substrate. Connecting traces on both sides. With the increasing complexity and density of components, many PCBs need to use both sides of the Printed Circuit Board.

The double-sided "PTH" (plated through holes) PCB is the universal workhorse of the electronics industry. In plated-through holes, copper connections go right through the connecting holes to the opposite side of the board. These PTH Connections either form simple electrical connections between both sides of the PCB (Via Holes), or electrical connectivity and mechanical support for leaded components. This makes the double-sided PTH PCB a much more physically robust item.

More flexibility for designers. Circuit density increased. Relatively low cost and Reduction of board size. Owing to their benefits, double-sided PCBs have covered a wide range of applications, including power supplies, industrial control, control relays, converters, UPS systems, LED lighting, hard drives, printers, mobile phone systems, power monitoring, test equipment, amplifiers, and traffic systems, like many other applications.

Multilayer PCBs

As the name implies, that is PCB boards with more than two layers, for example, four layers, six layers, eight layers, thirty-five layers, or even more. Many common features with double-layer PCBs, i.e., more than two layers' conductive traces, separated by insulating material between the layers, and the layer between the conductive traces connected through Vias, lamination as required

The advantage of multilayer circuit boards is that with multi-layer conductive wire and high-density drilling, the volume will be relatively small, and the weight will be relatively light accordingly. The high-density line reduces the space of components, which means more reliability. As there are more layers of circuits, it will be flexible for PCB design/layout.

Multi-layer printed circuit boards are the result of the development of electronic technology to high-speed, multi-function, high-capacity, small size. With the continuous development of electronic technology, especially the extensive application of large-scale and ultra-large-scale integrated circuits, multi-layer circuit PCBs with more requirements on fine line width, small

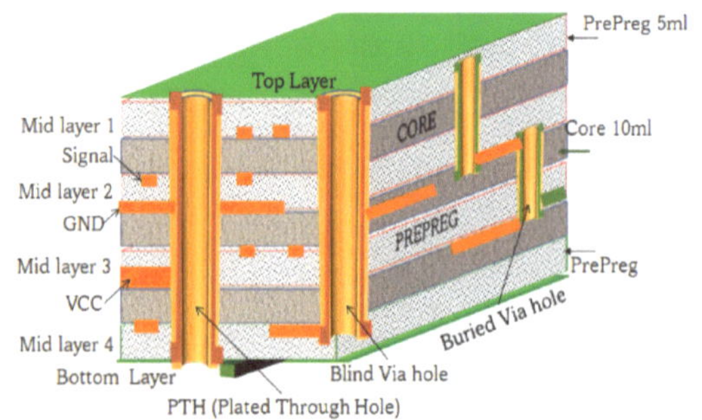

aperture through, blind and buried vias, high aperture ratio in developing high-density, high-precision, high layer to meet the needs of the market.

The advantages of Multilayer PCBs can be a reduction of board size and weight, a higher level of density and flexibility, and the capability of implementing multiple functions better at dealing with interference. It allows designers to produce very dense and highly complex designs. Quite often, the extra layers in these designs are used as power planes, which supply the circuit with power and reduce the electromagnetic interference levels emitted by designs.

Based on the merits mentioned above, multi-layer PCBs are applied in products asking for high technology and precision or those with higher space requirements, like satellites, computers, GPS technology, servers, data storage, signal transmission, X-ray equipment, hand-held devices, etc. Disadvantages – high cost and difficulty in testing.

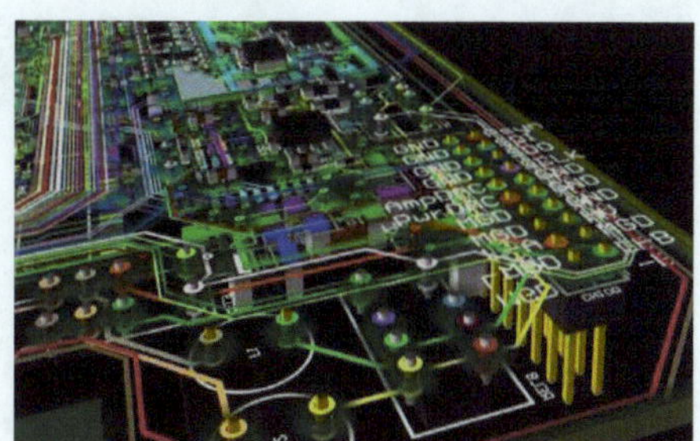

TYPES OF PCBS

1. Rigid PCBs:

Rigid PCBs refer to those whose base material is a type of solid material that cannot be bent, such as glass fiber. Rigid PCBs are typical PCBs now that they account for most boards.

Rigid flex PCBs

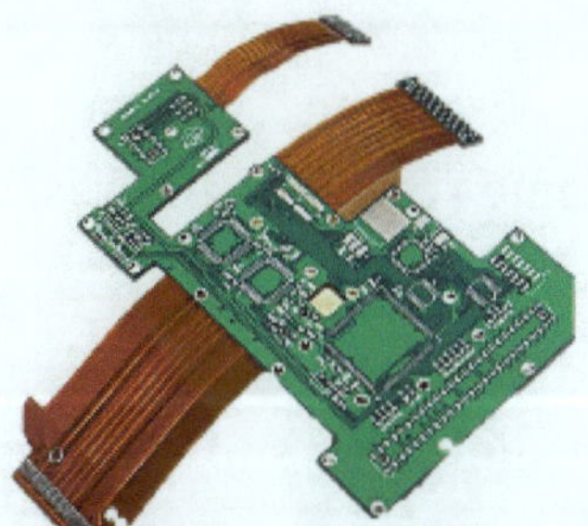
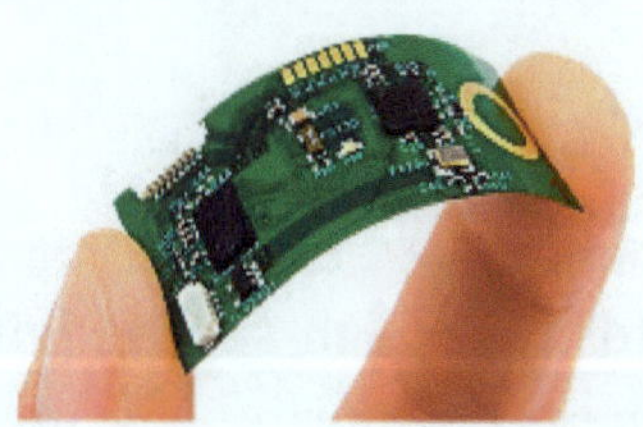

Boards using a combination of Flexible and Rigid Boards. By removing the need for connectors and cables between the individual rigid parts, the board size and overall system weight can be reduced. This type of process is applicable in cell phones, the Military, satellites, medicine, automobiles, etc.

Flexible PCBs

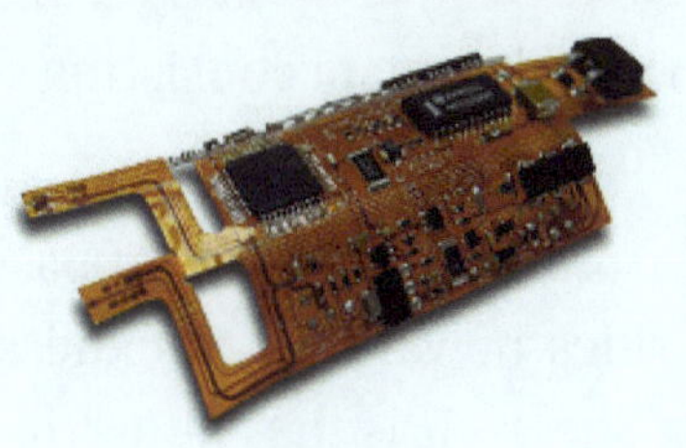

Assembling Electronic circuits by mounting electronic devices on flexible plastic substrates. Ideal for high and flex applications. Repair work is impossible for flexible circuits. Applications found in automobiles, disc drives, printers, etc.

Aluminum-backed PCBs

Typically used in high-power applications. It is the Ultimate solution for thermal heat dissipation and high-level mechanical stresses. This design keeps high-power components cool under heavy loads. Used in high-power LED products and switching power supplies.

BASIC UNDERSTANDING OF PCB SUBSTRATES

A Substrate, also referred to as Dielectric Material, is an insulating material sandwiched between two conducting layers of PCB. When the PCB is greater than two layers, a resin-impregnated cloth material, commonly referred to as Prepreg, is used as a substrate.

Glass-reinforced substrate, known as FR4, which is cheap, flame resistant, and resistant to absorption of water, is most used by PCB manufacturers.

For high-frequency circuit boards, i.e., High-speed digital and radio frequency designs, the substrate comes in mostly three types: 1. Fluorine – a high-cost dielectric substrate used for products with a frequency greater than or equal to 5GHz, 2. PPE resin and 3. modified epoxy resin- lowest price, usually FR4 used for products with a frequency range of 1 -10GHz.

Polyimide is a substrate that is capable of functioning in high-temperature environments and is highly resistant to fire. This substrate is highly recommended for Aerospace applications despite the water absorbent being a drawback, which is not relevant for this application.

PCB FABRICATION

The single-sided board is made from rigid laminate consisting of a woven glass epoxy base material clad with copper on one side of varying thickness.

Double-sided boards are made from the same type of base material clad with copper on two sides of varying thickness. Prepreg or Preimpregnated Bonding Sheet - It is the "glue "that holds the cores together. There are many types of materials, we use FR4 –a woven fiberglass cloth pre-impregnated with epoxy resin - known in the industry as B stage.

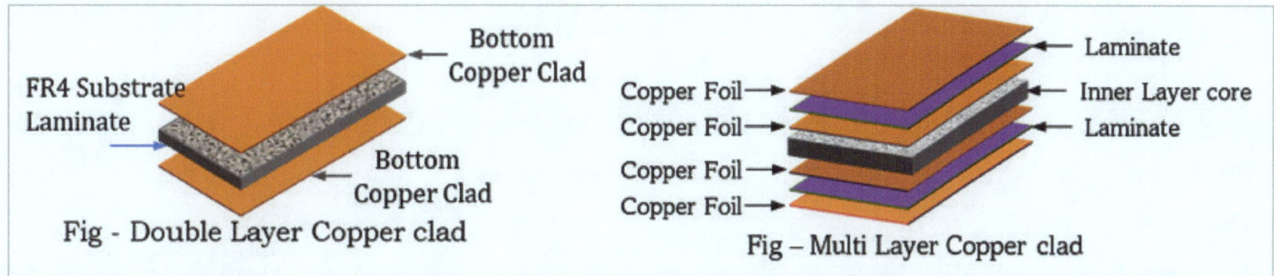

Multi-layer boards are made from the same base material with copper foil on the top &bottom and one or more "inner layer" cores. The number of "layers" corresponds to the number of copper foil layers. Multi-layer fabrication begins with the selection of an inner layer core –or a thin laminate material of the proper thickness. Cores can vary from 0.038" to 0.005" thick and the number of cores used will depend upon the board's design.

PCB MANUFACTURING FABRICATION PROCESS

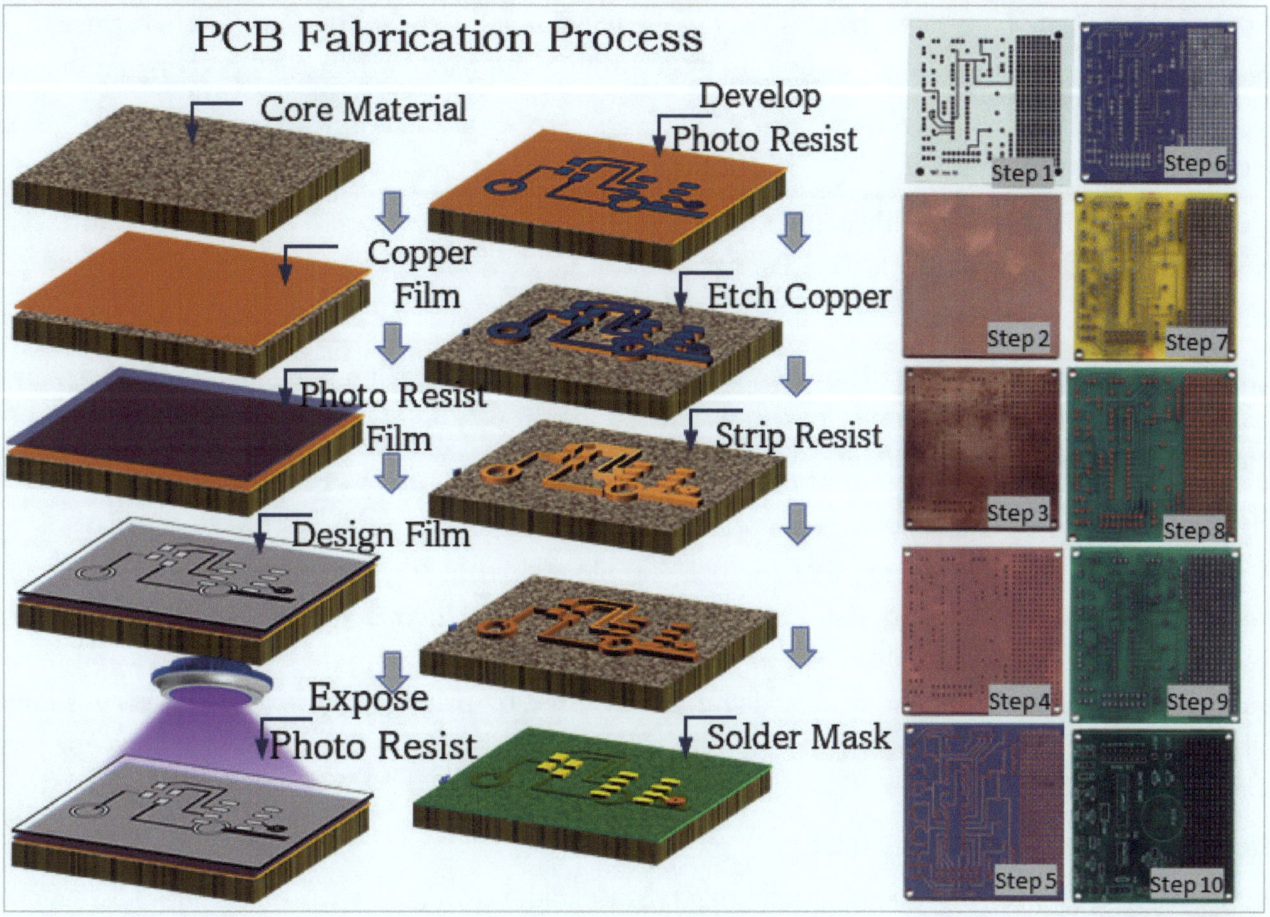

PCB Fabrication Process

Step#1 - Film Generation:

Generated from your design files, we create an exact film representation of your design. We will create one film per layer.

Step#2 - Shear Raw Material:

Industry-standard 0.059" thick, copper-clad, two sides. Panels will be sheared to accommodate many boards

Step#3 - Drill Holes:

Using NC machines and carbide drills, holes of various sizes are drilled through a stack of panels (usually 2 to 3 high). The board's designer determines the locations to fit specific components.

Step#4 - Electro Less Copper:

Once the smear is removed, a thin coating of copper is chemically deposited on all the exposed surfaces of the panel, including the hole walls. This creates a metallic base for electroplating copper into the holes and onto the surface. The thickness of the electroless deposit is between 45 & 60 millionths of an inch.

Step#5 - Apply Image:

Apply photosensitive dry film (plate resist) to the panel. Use a light source and film to expose the panel. Clear areas in the film allow light to pass through and harden the resist, creating an image of the circuit pattern.

Step#6 - Pattern Plate:

Electrochemical process to build copper in the holes and on the trace area. Apply tin to the surface. Note: All PCB express boards are plated through holes.

Step#7 - Strip & Etch:

The developed dry film resist is now removed from the panel. The tin plating is not affected. Any holes that were covered with resist are now open and will be non-plated. This is the first step in the common phrase "strip-etch-strip" or 'SES' process.

Step#8 - Solder mask:

A photo-sensitive, epoxy-based ink is applied, completely coating the panel. It is then dried to the touch but not yet finally cured. Using a method identical to the image, the panels are exposed to a light source through a film tool. Then the panel is developed, exposing the copper pads and hole defined by the artwork.   Solder mask is normally cured by baking in an oven; However, some fabricators use infrared heat sources. Apply the solder mask area to the entire board except for the solder pads.

Step#9 - Solder Coat: (HASL Finish)

Apply solder to pads by immersing them in a tank of solder. Hot air knives level the solder when removed from the tank.

Step#10 - Nomenclature:

Apply white letter marking using a screen-printing process.

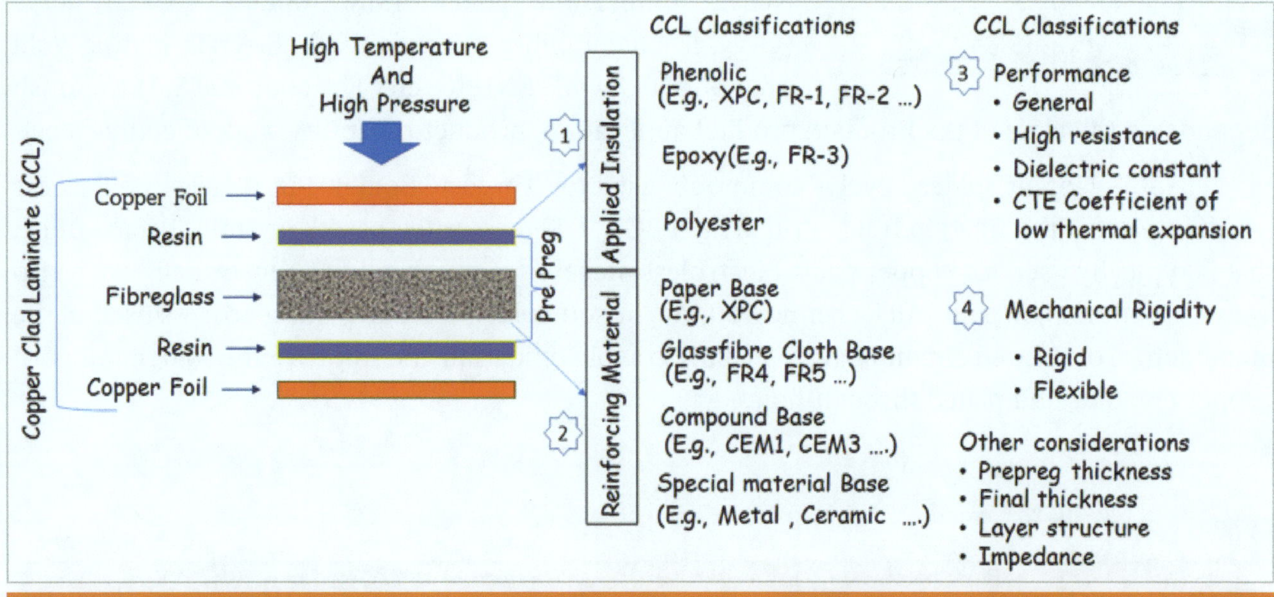

PCB PROCESS SURFACE FINISH TYPES

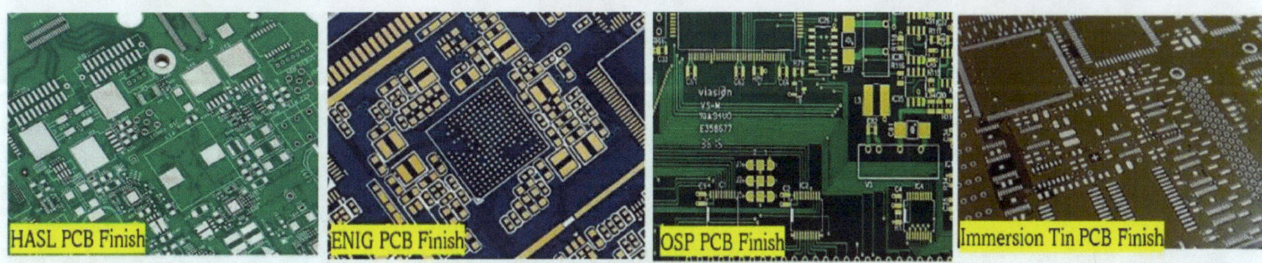

The surface finish forms a critical interface between the component and the PCB. The finish has two essential functions: to protect the exposed copper circuitry and to provide a solderable surface when assembling (soldering) the components to the printed circuit board.

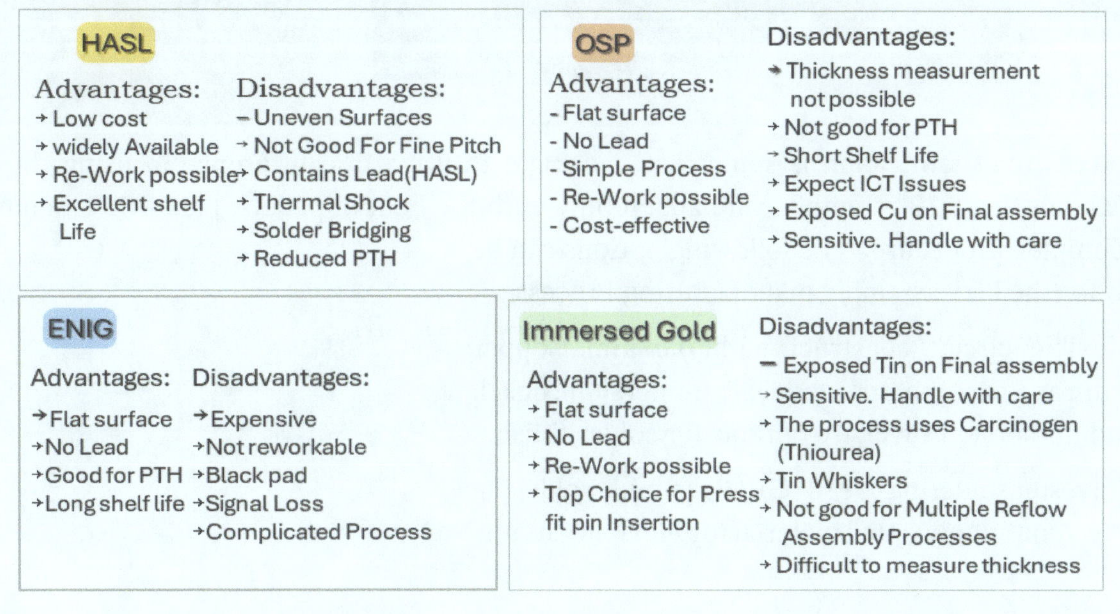

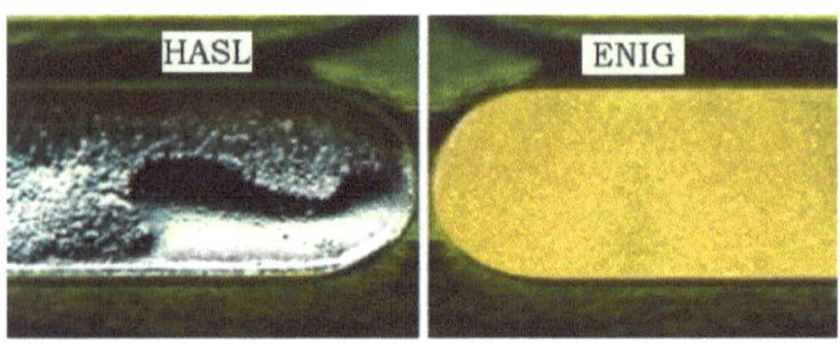

There are several types of surface finishes popularly known as HASL (Hot air Surface Leveler), ENIG (Electroless Nickel Immersion gold), IMMERSION SILVER, OSP (Organic Solderability Preservative), Electric plating gold fingers, ENIG+OSP, ENIPIG, etc. The choice of Surface finish depends on the class of the Product, product application, product reliability, and, of course, cost.

HASL – Hot air solder Level – commonly used for Tin Lead Solder. But it lacks the flatness requirement for the fine pitch Ball grid array FBGA. OSP is a water-based, organic surface finish that is typically used for copper pads. Electroless nickel immersion gold (ENIG or ENi/IAu), also known as immersion gold (Au), chemical Ni/Au, or soft gold, is a metal plating process used in the manufacture of printed circuit boards (PCBs), to avoid oxidation and improve the solder ability of copper contacts and plated through-holes

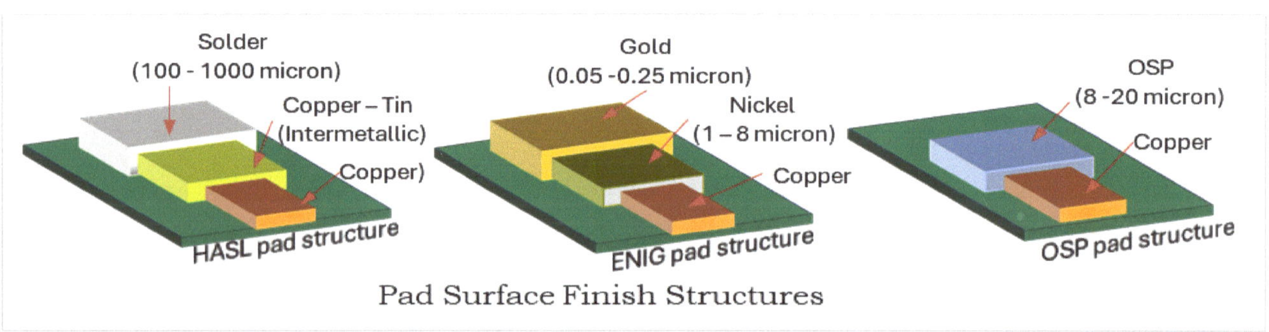

Pad Surface Finish Structures

1. Assembly of Flexible Circuits

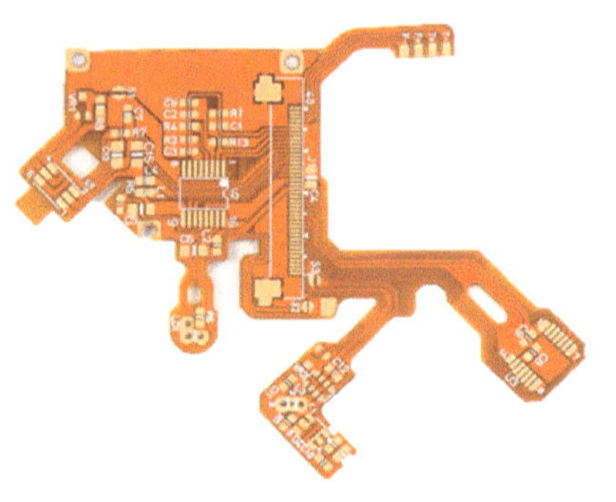
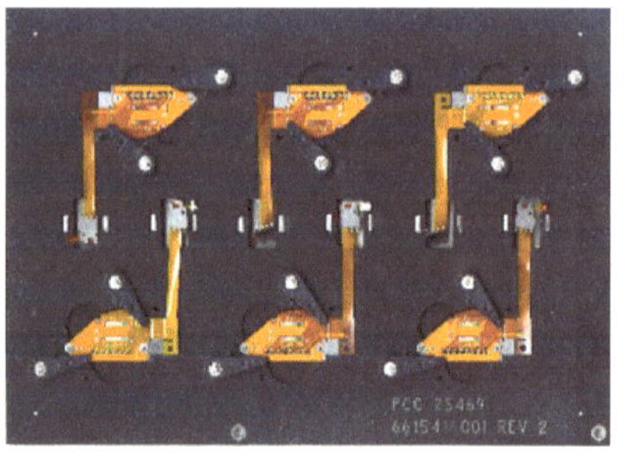
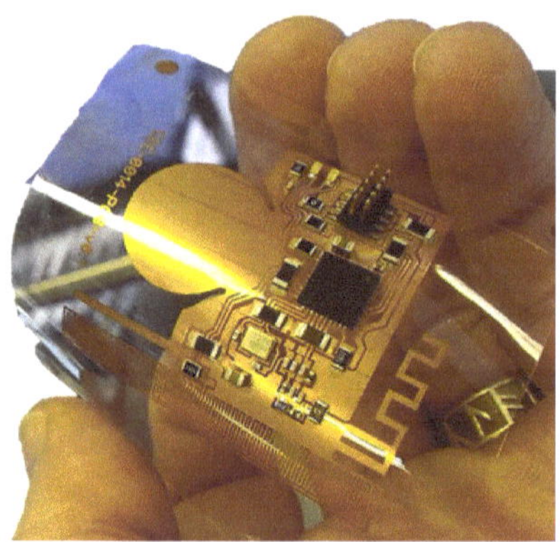

The Flux Circuit fabrication is somewhat like Rigid PCB, with a difference in using the Base Material, but flex PCB assembly can happen only with the help of pallets. Flex PCB assembly is. quite complex procedure. The following options can be chosen for the Flex assembly manufacturing Process.

Flexible circuit construct with 0.05mm Copper 18/18 µm as a base material and OSP, Immersion Nickel, Tin, and immersion silver are options for solder finish.

Polyester soldering (<150°C) / Polyamide soldering (<300°C) commonly opted material for Flexible Circuits

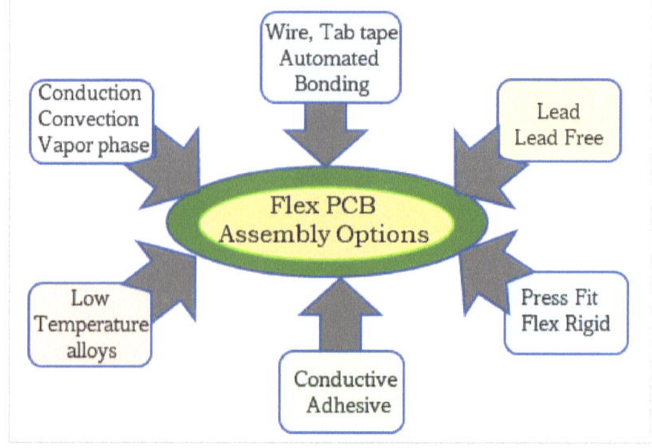

To prevent Warpage and to support a Flexible Circuit, use Kapton sticky tape on a corner, Pressure-sensitive Tact tape magnets, Clips, and Vacuum plates are the general practice for the assembly process. Electronic Manufacturing Industries normally follow IPC 1601 standards for Flexible PCB storage and baking guidelines

PCB RF TECHNOLOGY

Rapid advancements in RF/microwave low-frequency bands and usage have forced increased density of RF/microwave devices to achieve escalating frequencies to support the smaller, faster, and cheaper.

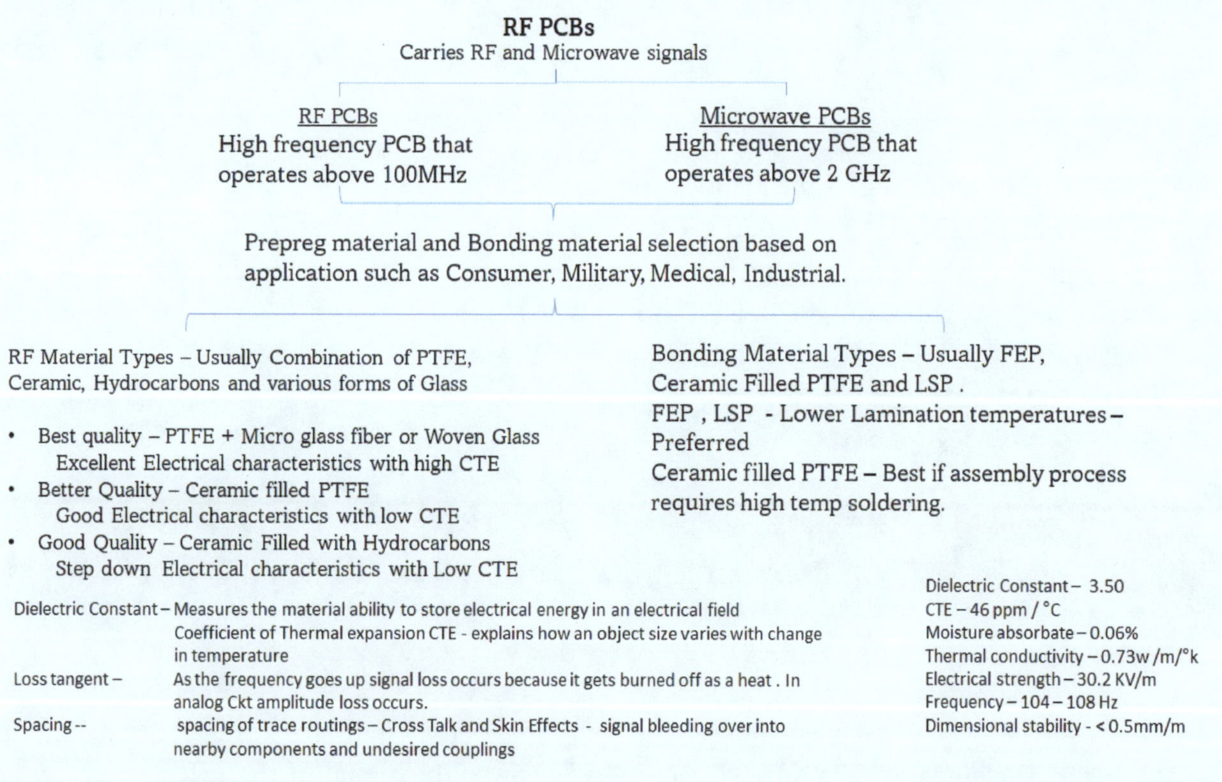

The design of RF PCBs is quite complex and needs to consider so many factors in all aspects because RF PCBs are the backbone of RF technology. It drives satellite communication systems, broadband access, and optical data networks. In addition, industrial sectors of automotive, industrial, military, homeland security, scientific, and medical applications are using RF technology to perform detection, measurements, and imaging functions.

Thermal management is the most important consideration that plays a very important role in the design of RF/microwave electronic devices. A large quantity of heat will be generated when signals are processed in high-frequency applications, particularly in the amplification of high-frequency signals. The reliable performance of an RF/microwave device depends on maintaining a constant value of the dielectric constant of the PCB's dielectric layer.

Edge Plating

Encapsulating the edges of printed circuit boards with plating may be required to improve EMI shielding of higher frequency designs and to improve chassis ground in electronic systems

Edge Plating

Cavity

Cavity designs can be applied in multiple locations at different depths on a single printed circuit board, and they can also be edge-plated. If the cavity is functioning as an RF/microwave resonant cavity, the frequency is determined by the size of the cavity and the PCB manufacturer must tightly control the X, Y, and Z dimensions of the cavity.

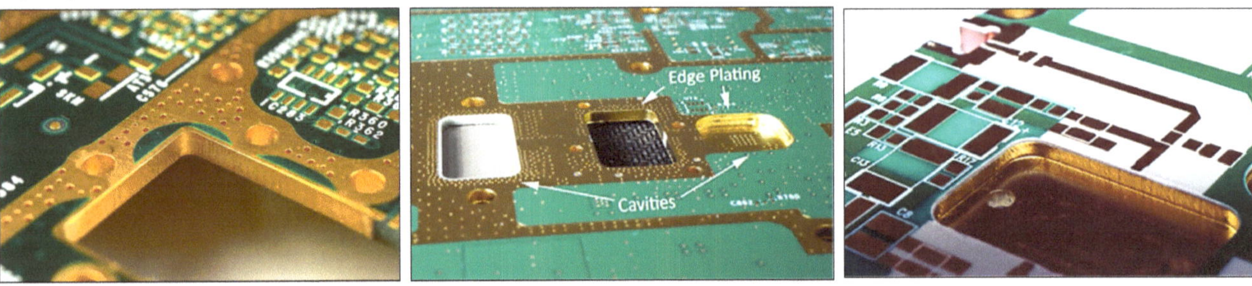

Cross Talk

Cross Talk is an unintentional electromagnetic coupling between traces on a printed circuit board. It triggers the signal pulses of one trace to overpower the signal of the other trace, even though they are not physically touching each other.

To reduce cross-talk, PCB designers configure the PCB layers in such a way that two adjacent signal layers with preferred routing directions cross each other, instead of running parallel to each other. For example, when layer number 2 is running "north to south," then layer number 3 shall run in the "east to west" direction. This method of design can minimize the possibility of broadside coupling.

Similarly, configuring ground planes between two adjacent signal layers can also reduce the chance of broadside coupling even more. At the same time, it provides a much better return path through the ground plane.

Hence, it provides much space or isolation between high-speed routing (differential pairs, clock routing, etc.) and other routing. It means space between traces at three times their line width, measured from center to center. This way, over 70% of their electrical field can be stopped from mutual interference.

PCB FABRICATION PROCESS DEFECTS

Issues arose during the lamination process. Issues with the laminated base material. Delamination is the most critical issue during PCB Fabrication. Moisture traps in the PCB base material. Manufacturing Process steps like the pressure of drilling Vias, through holes, or Routing could be the causes. Similarly, Line/track width issues like full or Partial etching. The other process defects illustrated below are related to the base materials for basic understanding.

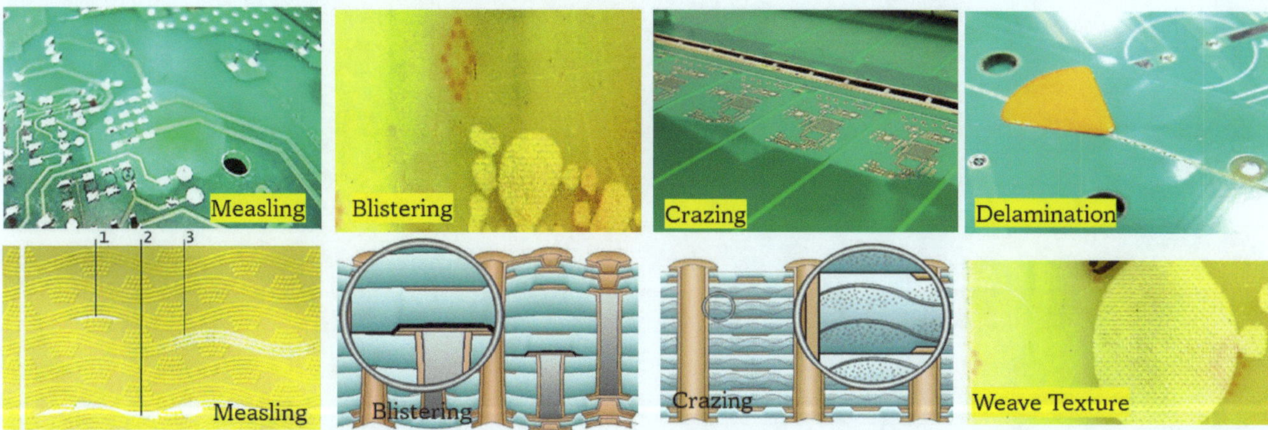

Measling is a form of delamination or separation in the laminated material. Crazing is a type of process defect where glass fibers are separated from the resin at the Weave. Blistering is another common quality defect that forms during the manufacturing process. Poor bonding force, Oil strains during drilling /milling/lamination, electroplating chemical treatments, leakage of the substrate due to excessive micro etching, etc., could be the Phenomenon of blisters.

IPC - A - 610 Standards define the acceptance criteria of these defects based on defect intensity. Quality requirements and acceptability of defect conditions are well defined in this standard for Class 1(general electronic products), Class 2 (dedicated service electronic products), and Class 3 (High Performance /Harsh Environment Electronic products) products.

SUMMARY

PCB design is the first step of Product development. Best PCB design is the key success of the product in terms of Manufacturability, Serviceability, and functionality. Choosing the best suitable substrates, choice of surface finish, and laying circuit guidelines concerning product application and class of product overcomes destructive interferences like leakage resistances, IR voltage drops in trace foils, vias, ground planes, the influence of stray capacitance, and dielectric absorption (DA), which influences bad performance. In addition, the tendency of PCBs to absorb atmospheric moisture

Another very broad area of PCB design is the topic of grounding. Grounding is a problem area for all analog and mixed-signal designs. PCB designers never mix high-frequency, high-current circuits with low-frequency, low-current sensitive circuits and are sure to separate the grounds. Even analog and digital components are separated physically and electrically, especially their grounds.

Further, the outstanding characteristics of PCBs must have thermal stability, resistance to oxidation, acids, bases, and other chemical agents like alkalis. It should have excellent dielectric properties and be Insoluble in water. Solderability, reliability, and reduced risk of moisture intrusion.

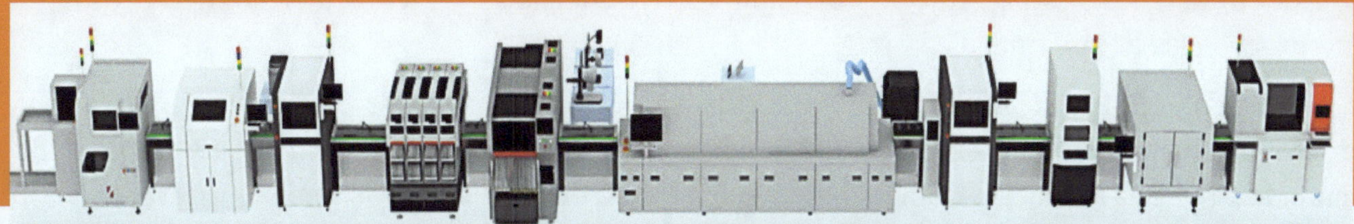

PCB Assembly Manufacturing Process

PREFACE

The manufacturing process for any industry can be defined as the collection of technologies and methods used to determine how the products will be manufactured. i.e., various process steps through which the Raw materials are transformed into products. The production process transforms the range of inputs into those outputs that the market or customer requires.

Depending on the manufacturing industry, whether it is the automobile industry, cement industry, Chemical industry, Petrochemical industry, Agriculture industry, Steel industry, Marine, Aerospace, Garment, or Electronic Manufacturing industry.... whatever manufacturing industry - depending on nature of the product, the adapted manufacturing processes, methods, and safety regulations can be varied. Ultimately, the production process transforms the range of supplied raw materials into specified products with mutually agreed product specifications. Agreed product specification means the output product can be either a finished product or a semi-knockdown product, which can be assembled in a separate manufacturing unit.

While raw material /work material transforms from one state to another state, the defined process operation changes the characteristics of the work material in terms of its shape, geometry, properties, appearance, etc. During this process, to get the desired product, the work material passes through various operations like material handling, transporting, processing, packaging, Inspection, testing, storing, etc.

The manufacturer follows the various relevant quality standards at each defined step to get the defined quality product. During the process, manufacturers make sure of environmental regulations, safety regulations, and waste management. Hence, the term 'manufacturing' refers to "to make or produce goods in large quantities using Machinery." Let us understand how the electronic manufacturing industry works.

ELECTRONIC MANUFACTURING – INTRODUCTION

Typically, you come across three types of electronic manufacturing industries. Those are TYPE 1, TYPE 2, and TYPE 3. **TYPE 1** Industries are those, let us say, Manufacturer-A, where all BOM material will be supplied by the respective suppliers. The supplied material undergoes Inspection before being issued to the production line from which the final product comes out. This final product undergoes Testing, adjustments, and quality checks as per company-defined processes and procedures before final packing.

Normally, EMS (Electronic manufacturing services) comes under this category. This final product for manufacturer-A can be considered as an SKD (semi-knock down) or subassembly for manufacturer-B. This subassembly would become a part of the BOM for the manufacturer-B for his final product. Hence, EMS companies take different orders of different products from their customers to make production. Well-known Companies of this type – Jabil, SFO Technology, Sanmina, Flextronics, etc.

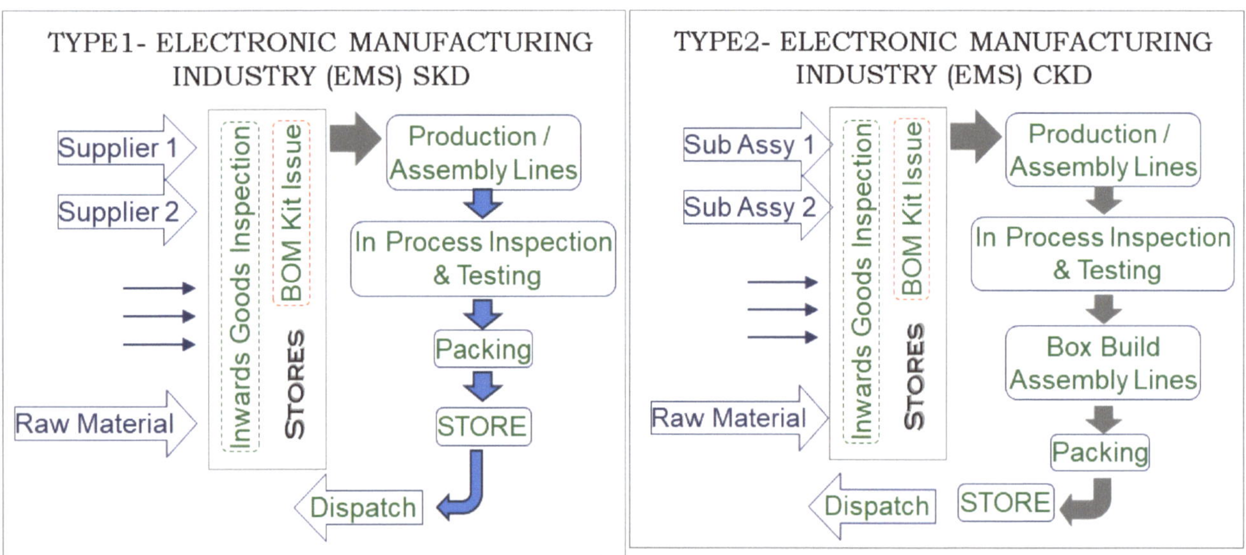

TYPE 2 Electronic Manufacturing Industries are those, let us say Manufacturer II, all BOM material supplied by the different supplier sources. In this process, BOM material could be raw materials, sub-assemblies, Electromechanical items, and cases /plastic boxes where the final product will fit.

In these cases, the final product from other manufacturers will become the supplier of the subassembly for manufacturer II. Finally, the BOM (Bill of material) material kit is issued to the main production line where the product undergoes various defined processes to get the final output, which in turn fits the box build operations.

Normally, this type of industry can be considered as OEMs (original equipment manufacturers), also referred to as CKD (Completely Knockdown) form products. In most cases, the final product goes to the direct user. Hence, OEMs can have their own branded custom-designed products, which are directly supplied to the user.

TYPE 3 Electronic Manufacturing industries are like Type II. The only difference is that most sub-assembly products are also manufactured in sub-assembly lines, and the final product of these will be absorbed into the main production lines as a part of BOM. Mostly, this type of industry can be considered as OEMs (original equipment manufacturers) also referred to as CKD (Completely Knockdown) form products. In most cases, the final product goes to the direct user. Hence, OEMs can have their own branded custom-designed products, which are directly supplied to the user.

The advantages of Type III processes are that the manufacturer can have good control over his processes and quality standards, and any design and manufacturing issues will be directly resolved within his premises.

Most TV manufacturing industries, mobile manufacturing industries, aerospace, and PC manufacturing come under this category. Sub-assembly's high voltage power supply assembly, Speaker preparation, heat

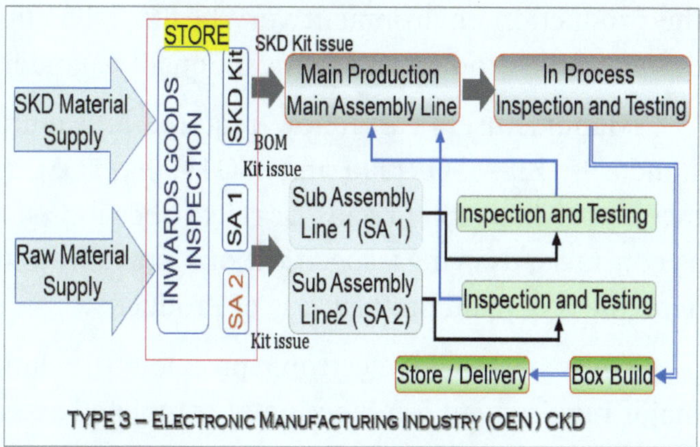

sink preparation for the TV industry, and sum assemblies like memory cards, power supply units for motherboards in PC manufacturing industries, and sub-assembly head track for Hard disk drives, etc. will be produced in separate assembly lines. These lines could be either automated or manual.

Whatever the case may be, whether it is OEM or EMS manufacturing, the product undergoes various processes, steps, and methods and comes across quality inspections and quality checks as per manufacturing process instructions during the production line.

There will always be control on parts quality, in process, and control test on the final product. During production and box build, can have various testing like electrical checks, alignments, calibrations, and tuning as per required specifications and incorporation of embedded systems e.g., Flash memory, patented software, etc. Control tests like drop tests, vibration tests, life tests, temperature tests, and safety tests will be done as per defined procedures to meet the agreed quality standards.

In the present era of business, demand for product supply has drastically increased, and consumer demands have as well. The business ethics of OEM and EMS completely changed, and new business deals such as CEM and ODM came into the market to cope with global market demands.

Original equipment manufacturers (OEM) typically focus on research and development and get Intellectual property rights for their products. Also, it offers their components and subsystems resold by another company as part of their product.

ODMs (original design manufacturers) are just like OEMs and EMS. They design, get intellectual property, manufacture, and sell their products. Typically, they specialize in a small number of specific products. Unlike OEMs, ODMs offer business to other companies through modifying/manufacturing their product in the name of another branded company (simply putting their Brand name) as if that product is designed and manufactured by that other branded company, and they sell through the requested source branded company.

The manufacturer may initiate approaches like Kaizen (Kaizen is a Japanese word that means 'continuous Improvement'), where working operators initiate group discussions within themselves and bring the solutions for process defects, Suggestion schemes, and Training to keep the production environment very healthy. The built-in quality of a product comes through the design of the product and stringent quality standards applied during the manufacturing process.

Manufacturers can follow required and mutually agreed standards most commonly - IPC standards, BIS, ISO 9000 and ISO14000, TQM, JIT etc. depending on the CLASS of product manufacturing. Accept and /or reject decisions shall be based on applicable documentation such as contracts, drawings, specifications, standards, and reference documents. The levels of quality agreement depend on the Class of Products.

Class I– General electronic products – Includes products suitable for applications where the major requirement is a function of completed assembly. That means no reliability.

Class II - Dedicated service electronic products – These include products where continued performance and extended life are required and for which uninterrupted service is desired but not critical. Typically, an end-use environment would not cause failures.

Class III – High performance / harsh environment electronic products – Includes products where continued high performance or performance-on-demand is critical. Product downtime cannot be tolerated, the end-use environment may be uncommonly harsh, and the product must function when required, such as life support or other critical systems. Defense products, semiconductors, the auto industry, and medical are normally categorized under Class III products.

The customer (User) has the ultimate responsibility for identifying the class to which assembly is evaluated. If the user and manufacturer do not establish and document the acceptance class, the manufacturer may do so.

TYPICAL ELECTRONIC MANUFACTURING PROCESS FLOW

A typical EMS flow chart is shown below for a Clear understanding of how the Manufacturing process instructions work at various stages, step by step. Before accepting any order, the manufacturer conducts contract feasibility with joint decisions by the Engineering, Quality, purchase, production planning, and production departments for relevant aspects of customer requirements.

These relevant various departments start working on the possibility of contract execution in all aspects with a clear understanding of the identification of resources (i.e. Equipment and its capability, training, and manpower), material planning and procurement, process capabilities, Manufacturing Inspection and Testing, release of Technical documents(Standard/product specifications) workmanship criteria, release of process specifications (setup charts, control limits, sampling plan) and finally pre-production acceptance run. Based on customer feedback, review, and corrective actions, the mass production launch happens.

A LITTLE ABOUT TYPICAL PROCESSES

Incoming material

All incoming material will be received at the store and inspected by IQC (incoming quality control). Material accepted and rejected will be sent back to the stores with proper Identification. Only accepted material will be issued by the store to production. Incoming material will be inspected either on sampling by using a Statistical sampling table or 100% Inspection. Acceptance and rejection of the lot will be based on the acceptance quality level. Material Inspection criteria can be defined through technical drawings, jigs, and fixtures.

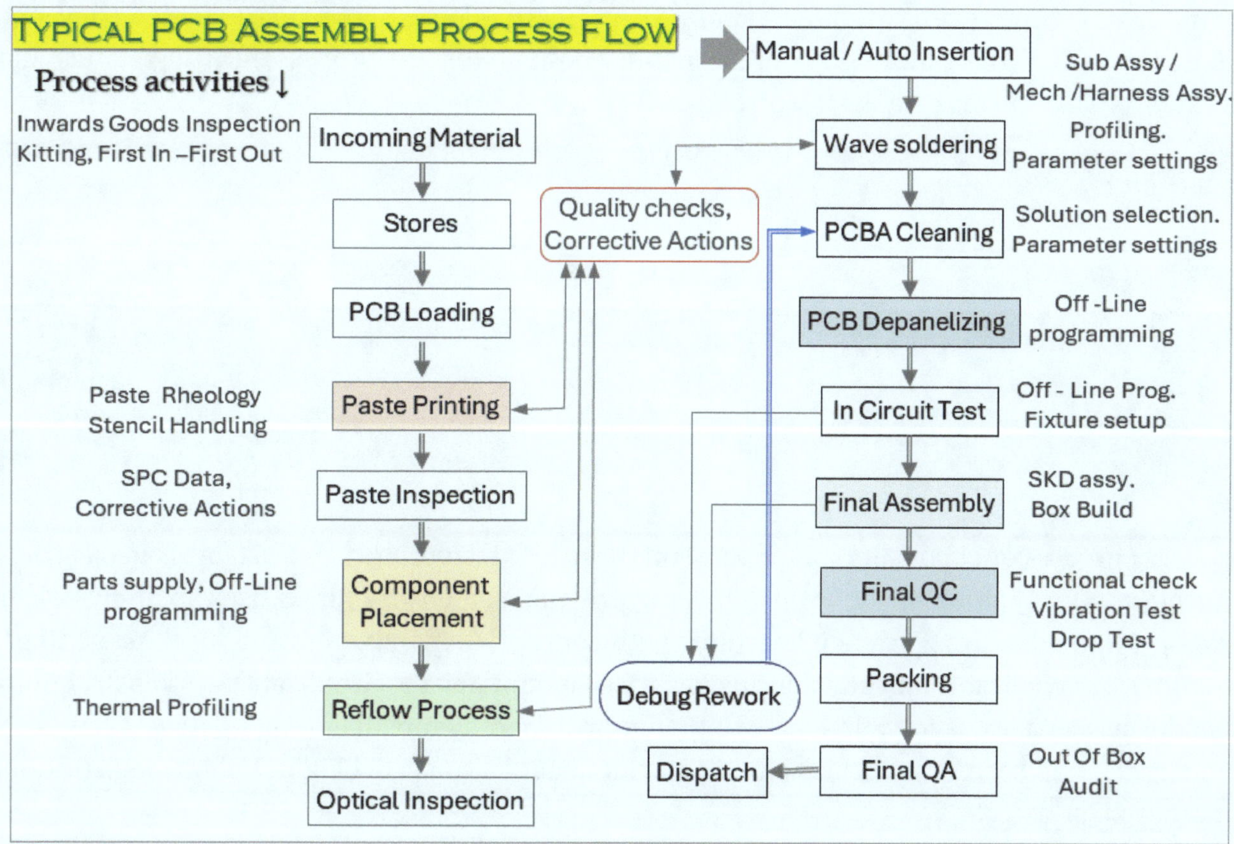

The Inward Goods Inspection department initiates the meeting MRB Material review Board with relevant authorities from Engineering, Production, Quality, Purchase, and Finance, and corrective & preventative action plans for about in-line rejections, incoming material rejected lots, either rejected in sampling inspection or out of specification rejected lots. MRB decides, based on mutual agreements, whether the rejected lots are to be reworked, used as it is, or rejected. In such cases, sometimes, based on supplier quality rating, the Vendor appraisal process will be re-initiated. A periodic Vendor Surveillance Audit was established and enabled periodic review of the quality management systems of approved suppliers.

Stores

The proper storage of material varies from manufacturer to manufacturer. Some use conventional methods, and some adapt modern technology concerning Industry 4.0. The storage of Shelf-life items and hazardous Materials will be carried out in separate, safe, identified areas. The FIFO (First In First Out) Concept will be followed. The status of the material, its identifications, and stored locations will be properly assigned.

The material received in HSC packing shall be verified as per work instructions. The HS (humidity sensitive) component baking instructions were issued to the assembly section of production, where components were baked for a specified time and environmental conditions before use. Baking temperature settings and time vary for the Tray components and the reel components.

The present technology changed the concepts of stores with the introduction of JIT (Just In Time), ERP (Enterprise Resource Planning), and Bar code scanning smart storage concepts. It is easier to find the status of material quantity, WIP (work in process), Kitting status, and traceability. That means complete control over WIP.

Paste Printing

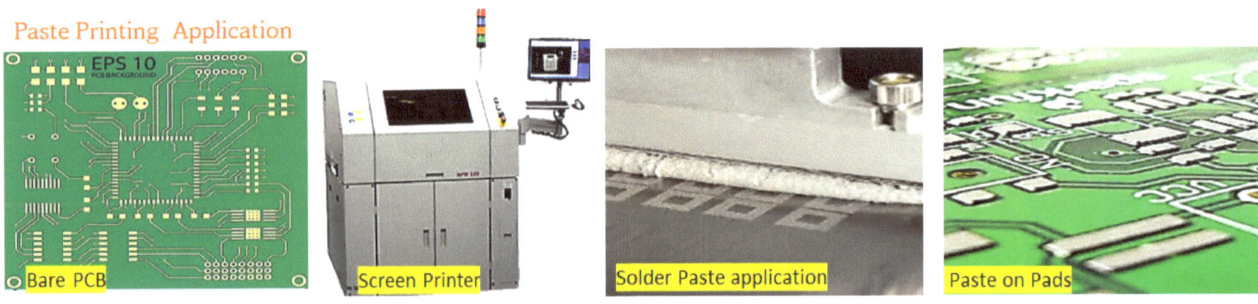

It is a process of depositing solder paste on the printed circuit board to establish an electrical connection after melting in the oven. The equipment, a screen printer, is used to apply solder paste on PCB pads through stencil apertures with appropriate parameters and machine settings. This process is applicable for the soldering of surface mount devices. Applicable work instructions on solder paste usage, Safety, defined equipment print parameters, handling of stencil, handling of squeegees, and corrective actions based on print quality, i.e., paste height, Volume registration, and other print defects.

Solder Paste Inspection

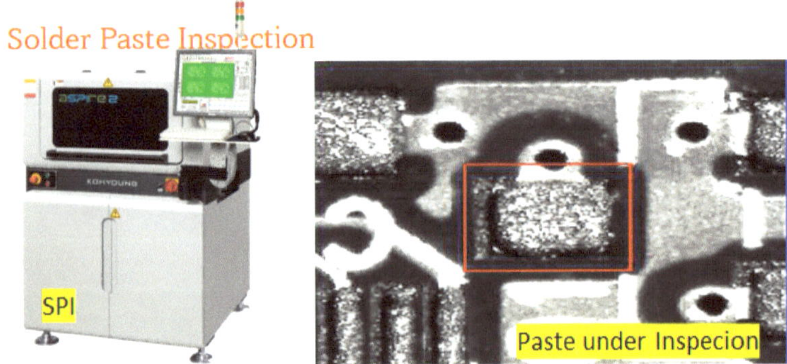

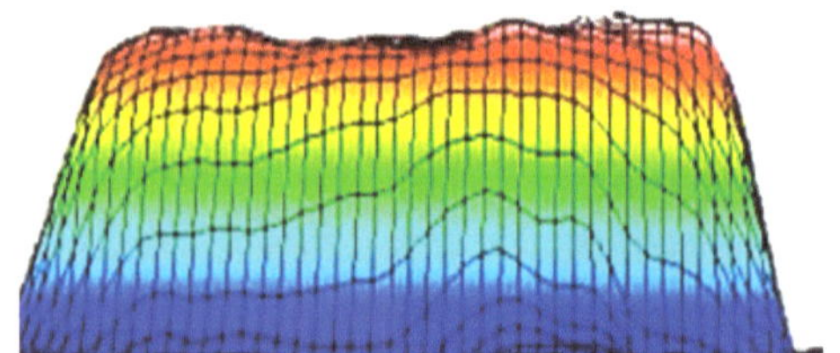

The purpose of SPI is to monitor the board on which the solder paste is deposited from the solder paste Printer. The purpose of this measurement machine is to detect abnormal patterns developed by the paste printing Machine (Screen printer) either due to the Printer defects or process defects.

The SPC software can correlate cross reference and analyze the data in real-time giving immediate feedback on the printing process i.e. Defect location on each PCB, Top 10 defects, real timer monitoring (line performance), 3D defect images, Cp/Cpk and various detailed Pareto and Pie charts which allows or signals process and quality engineers for preventive measures.

Component placement

Component Placement Process

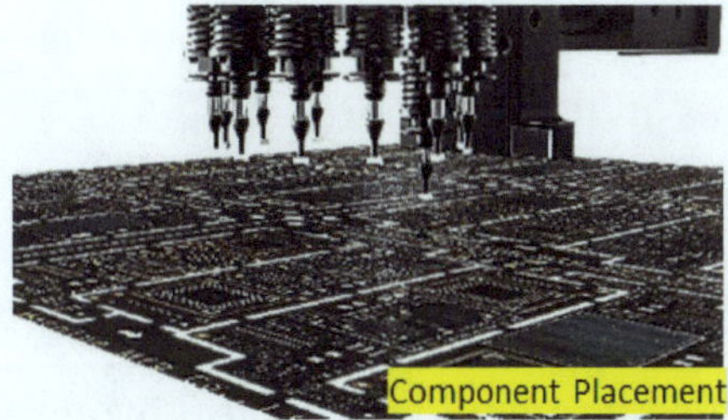

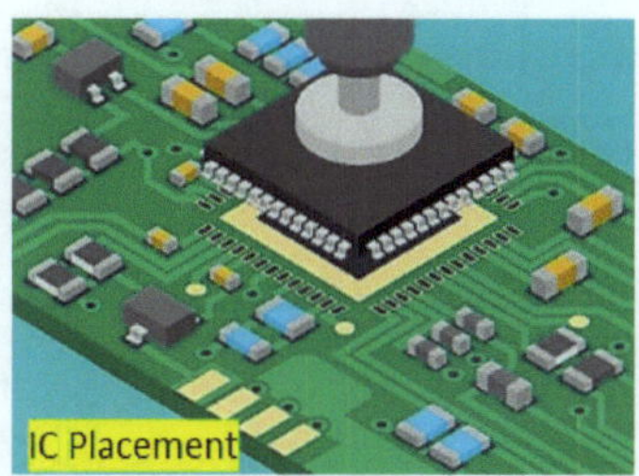

Pick and place machines are used for this process. As its name implies, it picks up the SMD components of a variety of packages with the help of appropriate vacuum nozzles and places them in the specified locations on the PCB pads very accurately with specified parameter settings. Paste-printed or glue-printed PCBs are loaded into the pick-and-place machine through a conveyor, and components are picked from specified feeding locations and placed on PCBs in specific locations as defined in the placement programs. Then, these populated boards are fed into reflow equipment to make intermetallic bonds through the solder-melt process.

Standards like IPC 610F (Acceptable criteria of electronic assembly) and Work Instructions on component loading, feeder loading, program loading, and use of Humidity-sensitive components will be followed. Procedures on dropout components will be defined. Improvements and corrective actions will be taken based on inspection feedback.

Reflow Process

Reflow Process

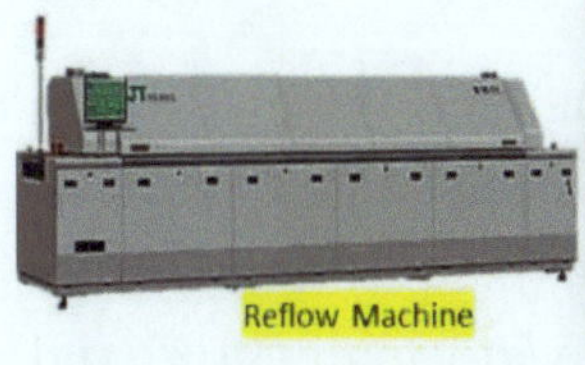

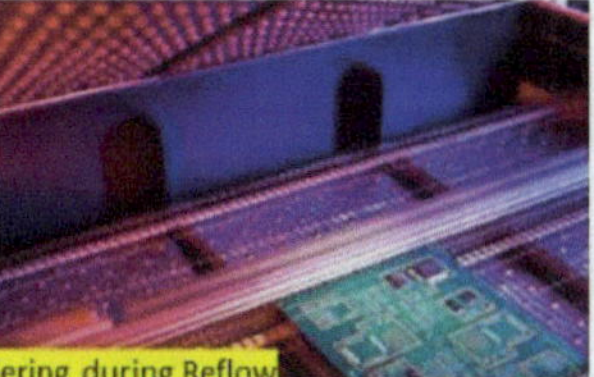

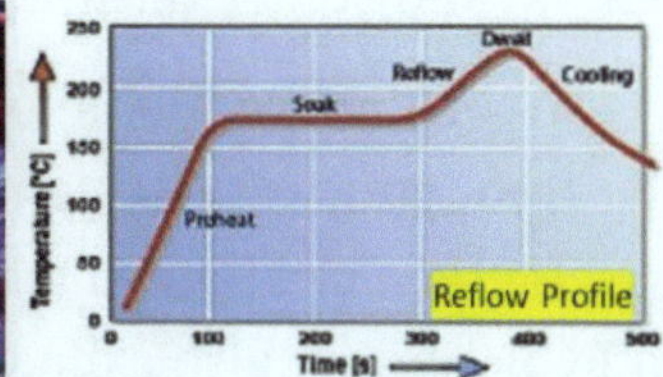

Populated boards fed into surface mount reflow equipment to make intermetallic bonds through the solder melt process. The manufacturing process instructions include the temperature profile of the PCB, temperature zone settings, conveyer speed, PCB orientation, and grouping.

A Thermal Profile will be carried out for each product change. Thermal profile records will be maintained. Post-reflow inspection criteria normally define 100% inspection for solder quality either by utilizing manual inspection through a 30X microscope or with AOI (automatic optical inspection), which is the ultimate solution for getting real-time process control feedback. Data collection based on profile, Height, and alignment, SPC initiated at this stage.

Manual Insertion

Manufacturing process instructions include Visual aids to assist the operator in the placement of components, bin location, and part numbers. Arrangement of Bins is one of the arts that reduces the operator's fatigue and mishandling of components at the same time, Bin arrangement methods improve the work speed. Visual aids will be placed in front of operators to indicate parts to be installed, the orientation of components, and any special features to observe. Work instructions also indicate some of the key components, such as "do not use dropped crystals and oscillators." PCB layout templates are used as an inspection tool.

Wave Soldering

PCB populated with Through Hole components and Glued SMD on the bottom of PCBs passed through this wave soldering Machine to solder all the Through Hole leads and Glued SMD leads instantaneously. Unlike the reflow Oven, instead of solder Paste, Molten solder is applied to solder the joints. Populated PCBs passed through a wave of molten solder. Hence, the process is called wave soldering.

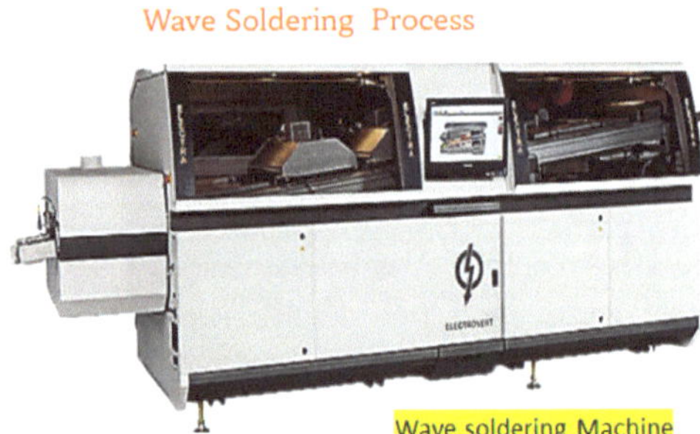

Process instructions include a profile of the PCB, Temperature zone settings, and Conveyor speed. Also, process instructions provide information on flex-specific gravity, solder pot temperature, solder composition, Wave type, PCB minimum spacing, PCB orientation wave height, and solder fixture orientation, which will be defined for each product.

Work instructions for the thermal profile maintained where it contains the information on the frequency of getting profile data, PCB part number, thermocouple locations, and all critical parameters, i.e., Rate of temperature change, solder pot temperature, and machine settings. Design of PCB Fixture, also called pallets with flexibility of orientation into the wave (normally 90°) and connectors, other components are held in the correct position while being wave soldered.

Inspection criteria will be defined for PCB solder quality, PTH component placement and alignment, and solder joint-specific requirements. A hundred percent inspection is mandatory in most applications, so AOI, Automated Optical inspection, is mandatory in the present Assembly Process. Statistical process control SPC will be set based on inspection reports.

PCBA Cleaning Process

Aqueous Cleaning Machine

PCBA Cleaning

Assembled PCBs after the Reflow or Wave soldering process are undergoing a complete cleaning process. Though this is not mandatory for PCBA assembly, this process is an exceptional option for sophisticated Class III products and is mandatory when any production process involves Conformal Coating.

PCBA cleaning removes dissolved contaminants and suspended particles (solder Balls) and makes the board look clear. Functional test results will be good, as there are clear contact points. After cleaning, guarantee the reliability of electrical function and the lifetime of electronic products. The reason OEMs and EMS value cleanliness.

Manufacturing process instructions include cleaning system parameters that provide information on solvents (i.e., type of solvent, change frequency, acceptance of contamination level, specific gravity), conveyor speed, nozzle pressure, nozzle angle, and maintenance schedules. Cleanliness can be checked with an Omega meter or a logograph.

Rework & Repair

Manufacturing process instructions include solder iron temperature and checking frequency, Instrument calibration frequency, type of flux, solder wire chemical composition (lead or lead-free), ESD requirements, and special instructions, if any.

Rework and repair manufacturing process instructions include temperature profiles on all component types, protection methods for surrounding components, and the size of nozzles for different part sizes. Since rework and repair is the most complex process, suppliers must train and certify operators in part replacement and rework.

Conformal Coating Process

Some applications such as defense, Aerospace, marine, energy, automobile, or electronic products are going to be exposed to harsh environmental conditions such as extreme weather, dust, humidity, etc. Hence, the protection of PCB assembly or circuitry inside the product is very much essential despite good design and quality. It should be weatherproof, shock and vibration proof, which means products should be protected from all kinds of extreme environmental conditions...

In other words, shielding electronics from moisture, chemicals, and contaminants, handling and abrasion, temperature extremes, and radiation.

Either Potting or Conformal coating and encapsulation resins are one of the best solutions for the protection of electronic circuits. Both Potting and Conformal coating processes protect the electronic circuit board, but a selection of processes during the production line lies with product application and the degree of environmental protection required.

Potting and encapsulation resins offer the highest level of protection for PCB assembly. Resins applied 0.5mm coating thickness or over. Conformal coatings are generally thin films applied in the range of 25 microns to 250 microns, coating dry film thickness range. The value of Coating thickness is always considered in a dry state, and often, coatings are clear, and coated components are visible. Most coatings are single compound materials, and, in some cases, two or three compounds are mixed to apply for coating purposes.

An Aqueous Cleaning process is mandatory before the conformal coating process because once the coating has delaminated, moisture under the coating layer eventually collects and forms an electrolytic solution. Even for no-clean flux cases, conformal coating states that no clean flux residues can reduce adhesion, potentially resulting in delamination and micro condensation.

In-Circuit Test

In Circuit Testing (ICT) is one of the important assembly processes in manufacturing assembly lines and is considered a manufacturing verification tool in the electronic manufacturing industry. It tests the individual components; the Component is the connection with the PCB and, to some extent, the product functional test.

ICT is a powerful tool for printed circuit board testing. The purpose is to find both component and manufacturing defects before the assembly is completed. ICT is used for both BBT (Bare board test) and after assembly. It is possible to undertake a very comprehensive form of printed circuit board test, ensuring the circuit has been manufactured correctly and has a very high chance of performing to its specifications.

ICT carries BBT for PCBs to find shorts, open, trace cross talk, and in assembly phase - measures the values of components like Resistors, Capacitors, inductors, Diode or Transistor, Zener diodes, DC voltage, Short or Open measurement, Transistor with built in resistor, FET and Opto couplers.

1. The capability of measurement range for each mentioned component differs from ICT manufacturer to manufacturer and model to model.

Types Of In-Circuit Testers

1. The main types of ICT machines that are available and the requirement type depend on the manufacturing or test process used, the volume, and the Boards that are used.

1.   Standard ICT Machine - Generally called IN Circuit Tester. Widely accepted test machine

2.   Flying Probe Tester - This has a simple fixture to hold the board, and contact is made via a few probes that can move around the board and make contact as required. These are moved under software control so any board updates can be accommodated with changes to the software program.

3.   Manufacturing defect analyzer, MDA -   This form of tester offers a basic In-circuit test of resistance, continuity, and insulation. It is just used for the detection of manufacturing defects like short circuits across tracks and open circuit connections.

4.   Cable form tester: This form of tester is used to test cables. It uses the same basic functions as an MDA, although some form of high voltages may need to be applied occasionally to test for insulation. Its operation is optimized for the testing of cables.

ASSEMBLY PROCESS TYPES

The electronic manufacturing industry adopts various types of processes for their PCB assembly production, depending upon the nature of PCB assembly desired for the Product as per the product design and applications.

1. Single-side through-hole component Mounting

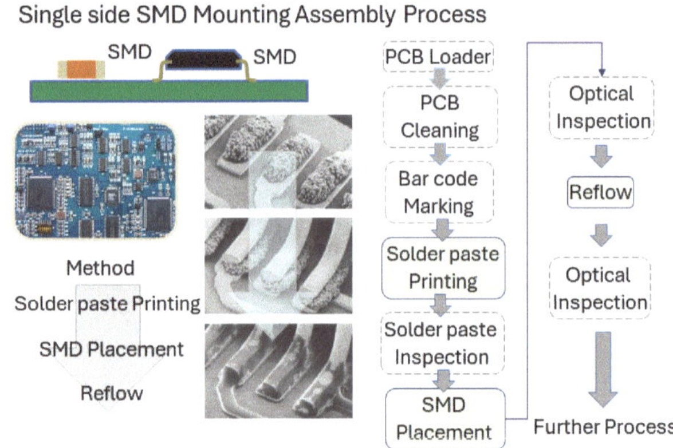

Only Through Hole Components are designed to assemble on a single side of the PCB and to solder circuit joints. This process does not require a Paste Printing and reflow process. The typical manufacturing process model is shown below.

PCB Loader, THC auto Insertion, MI Line (manual insertion), and wave soldering processes are mandatory while the rest of the process can be as per the quality Requirements.

2. Single-Sided Surface Mount

This type of board design only allows SMD components to be mounted on one side of the PCB and reflowed to solder the components to make interconnections. The process of Paste printing, SMD component placement, and reflow of the solder mounting are mandatory, whereas the rest of the processes can have importance depending on quality requirements. X-ray is involved in the inspection of leadless BGA components.

3. Double Side Surface Mount Assembly

The best example of this type of PCB assembly is Mobile PCBs. Only SMD components are populated on both sides of the PCB. The production assembly process is like a single-sided surface mount but requires a repeat process for side A after the completion of side B.

This type of PCB is normally designed in such a way that lightweight active and passive components with less population are on Side B and all active and passive sophisticated SMD larger packages are on Side A. Hence, the assembly process recommends the first reflow of Side-B and flip board for the repeat process of Side-A so that components of Side-B will not fall during the second reflow process. For further protection of components in Side B, dot matrix adhesive was applied in between the solder pads so that components do not fall during the second-time reflow process.

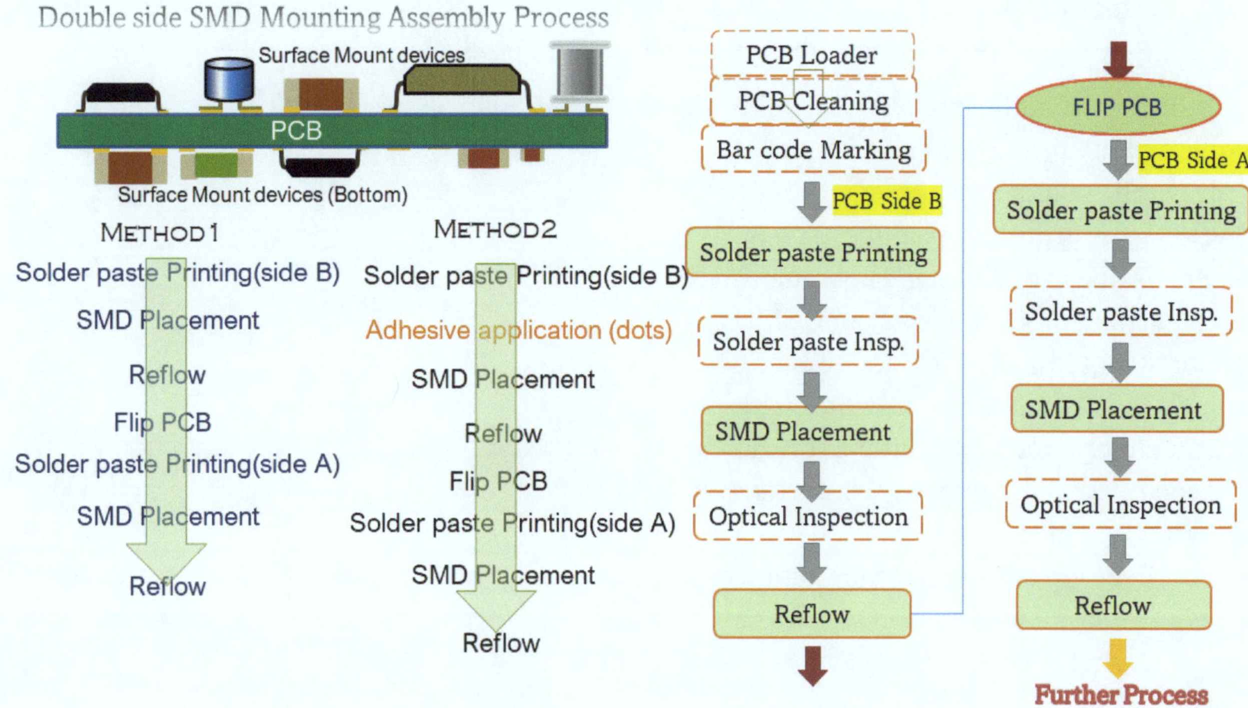

The processes of Paste printing, SMD component placement, and reflow of the solder mounting are mandatory, whereas the rest of the processes can have importance depending on quality requirements and the Class of the Product. X-ray is involved in the inspection of leadless BGA components.

4. Single Side Mix Component Mounting Assembly

This type of board has both SMD and THC packages assembled on one side of the board. PCB design of this type is quite common for many products. The PCB assembly process is complex for production lines as it involves all special processes like screen printing, Reflow, and Wave Soldering Processes.

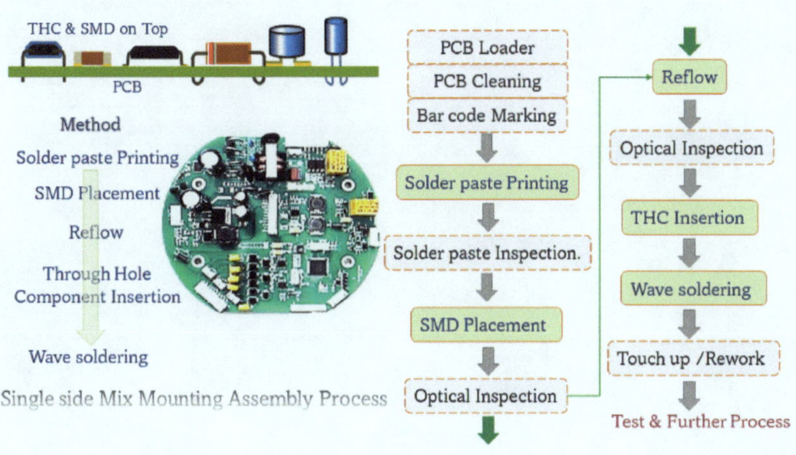

Additional mandatory processes, Through hole component Auto Insertion (Cut THC leads, insert in PCB at location and clinch from Bottom), along with Manual Component Insertion process and Wave soldering process, were added for soldering SMD and THC joints.

5. Double Side Mix Component Mounting Assembly

These types of boards have both SMD and THC packages assembled on the primary side of the board (normally called Side A of PCB) and only SMD packages on the secondary side (side B of PCB). PCB design of this type is quite common for many products

PCB assembly process is quite complex for production lines as it involves all mandatory processes like Solder Paste printing(Machine screen printer), Adhesive Printing (Equipment Screen printer or Glue dispenser), SMD placement(Equipment Pick and Place), Through hole component Insertion with Cut and clinch process (Equipment Auto Insertion), Manual component insertion Process(MI Lines), Adhesive curing process (Equipment Curing Oven) Reflow Process (Equipment Reflow Oven) and Wave soldering Process(Equipment Wave soldering).

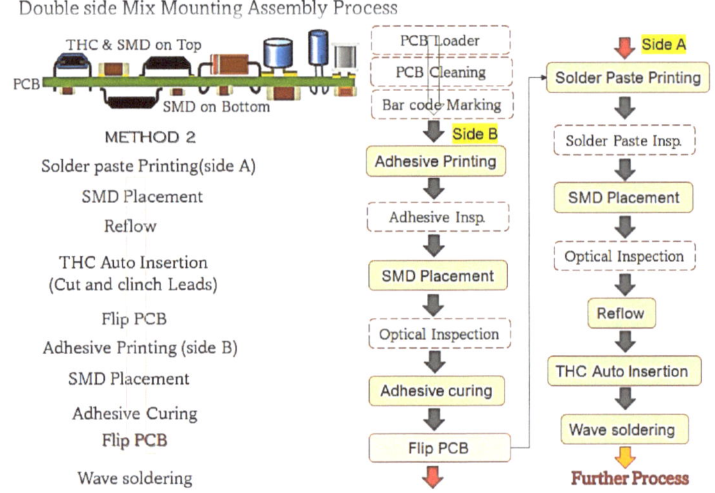

This type of assembly can be done in two methods with limited process differences. In the method 1 process, the application of Adhesive can be done either with a stencil or with Glue dispense, whereas in method 2, the application of adhesive is only possible with a Glue Dispenser.

Typical Class III product Double side Mix Mounting Assembly Process

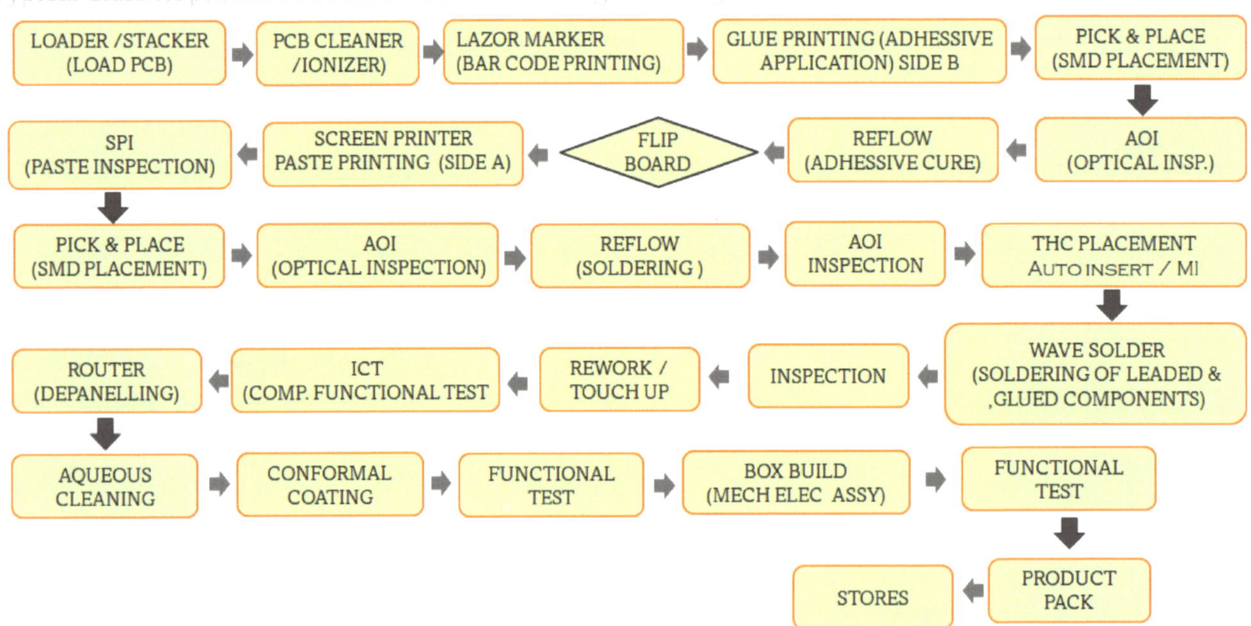

SUMMARY:

Present technology assembly lines, moreover, are built with inline inspection and measuring systems and product traceability through barcode/IR code scanning. 3D SPIs (solder paste inspection), AOI (Automatic optical inspection), and advanced BGA rework stations. Additional processes, conformal coating, laser marking, and x-ray inspections were incorporated to strengthen quality requirements. Embedded software like Fuji-Trax (OEM FUJI), Valor, ERP, etc., was introduced, which provides all information on SPC data at every stage of manufacturing, which ultimately avoids manual intervention and meets Industry 4.0 objectives.

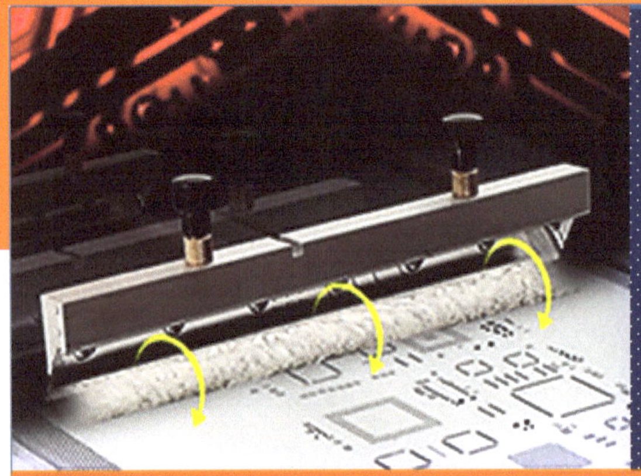

Screen Printing Process

Equipment Stencil screen printers in Surface Mount Technology play a crucial role as it has been well observed that the Solder paste printing and reflow processes cause 70% of solder defects. Ultimately, the PCB assembly main process starts with the loading of bare or flipped one-side assembled PCB into the printer, the clamping system or snugger and vacuum holder system are used to secure the X and Y axis of the board. The internal vision system aligns the stencil to the board, after which the board is attached close to the Stencil from the bottom with zero gaps. The print head moves, and the squeegee prints the solder paste. The stencil and board are then separated with proper snap-off parameters and unloaded. Optional feature: post-print inspection performed with a special 2D vision system on the printed board.

Though the process seems to be very simple, machine capability is one of the most important considerations when choosing the appropriate printer to build the board consistently. For fulfilling the objective of good paste printing, machine capability plays an important role despite taking care of solder paste specifications, stencil adherent to standard design rules, and PCB pad.

Equipment manufacturers provide optional features like a Temperature control unit, 2D Inspection for shift and area coverage, top clamping as a standard, optional side/snugger clamping, and custom-made vacuum clamping, which requires a dedicated work holder. For PCB support, optional tools, Gridlock, and flex gel, which are normally used for double-sided PCB assembly, backup pins are provided as a standard tool. Other optional features - Bar code reader for traceability and product changeover, automatic solder paste dispenser, paste rolling height check which controls paste dispense on stencil, and SPI interface tool for closed-loop corrective action when print shift defect occurs.

Alignment accuracy between the stencil apertures and the PCB pads is the key factor in getting consistent print volume with low standard deviation and high CPK. This factor is more critical when 01005s, BGAs, CSPs, and QFNs become more mainstream.

The fill process, i.e., the transfer efficiency of solder paste, depends on stencil aperture design, type of squeegee selected, and type of Solder paste chosen for the application. The paste transfer process through the stencil to the PCB pad also depends on the stencil wall quality and the aspect ratio of the aperture. Apart from these controls, equipment teaching parameters such as Squeegee down stop, squeegee speed, Squeegee pressure, and PCB separation speed from the stencil.

Hence, printing equipment must have the capability for controlling the parameters to maintain the correct combination of print speed, pressure, snap-off, under stencil cleaning frequency, and most importantly, vision alignment. Combination of print speed, pressure, snap off, under stencil cleaning frequency, and most importantly, vision alignment.

EQUIPMENT PRINTER (OVERVIEW)

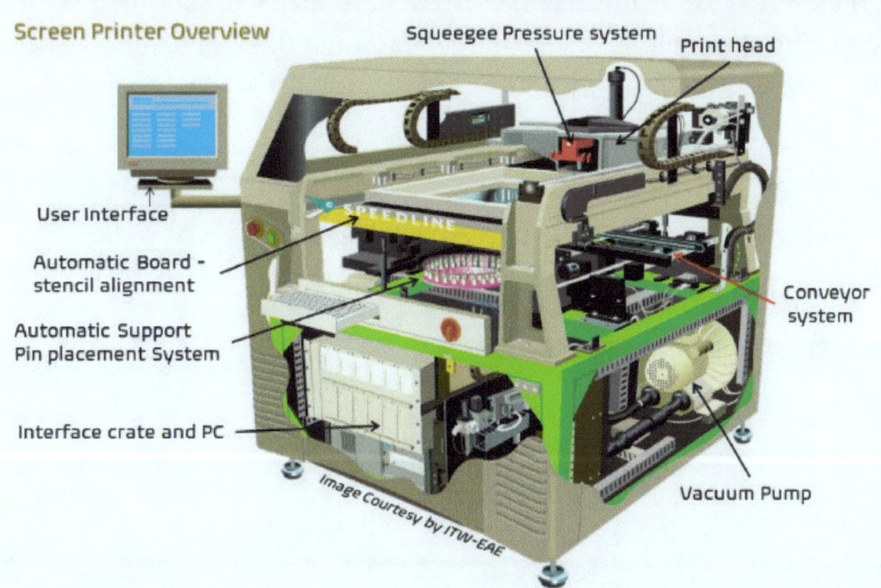

Let us consider this printer to understand machine overview and programming concepts. The printer can be equipped with the most common components – Print Head, Conveyor section, PCB clamping system, PCB Supporting system, User interface, SPC, vision system (Board to stencil alignment), vacuum stencil cleaner, and High and Low voltage circuitry.

Manual printers are low-cost printers. The quality of printing purely depends on operator skill. Reliability and repeatability are the main concerns, so it is not recommended for high-quality paste printing. So, Manual printers are only Suitable for Class 1 product manufacturing with fewer volumes and prototype production.

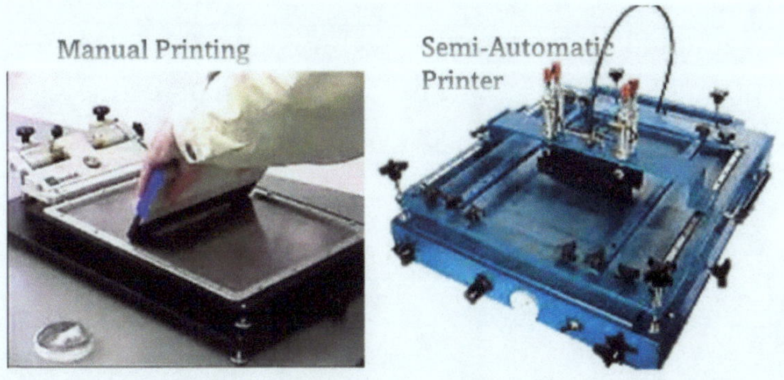

Semi-automatic printers are also low-cost printers. Manual skill is required for initial alignment and offsets, or tolerances cannot be adjusted automatically; most printers call for manual vision alignment. High frequency of mandatory inspection. Not recommended for fine pitch. Hence, these semi-automatic printers are ideal for prototypes.

Automatic printers offer automatic vision alignment and accurate deposition of solder paste on the pad, possible. Offers closed loop environment.

SOLDER PASTE STENCIL PRINTER – PRINT HEAD

The print head, a section of the printer, is attached to the x-axis and the y-axis robot. A Squeegee unit is attached to this print head, which moves forward and backward with the help of the XY robot and controls squeegees up and down during paste printing as per user set parameters in the Program. Depending on the printer variant, the solder paste auto dispenser unit and stencil stopper unit can also be a part of the Print heat section.

The print head controls three functions: the Auto calibration system, auto paste dispensing, and the type of controlled squeegee print head. Parameters – Print speed, print pressure/squeeze pressure, and Stroke can be defined either by manual entry or by Auto entry if equipped with these options. Enough paste supply onto the stencil in time and maintaining the role height is also the most important factor in solder paste printing to avoid less solder and dry solder. Hence, the auto-paste dispensing option is preferable to manual supply.

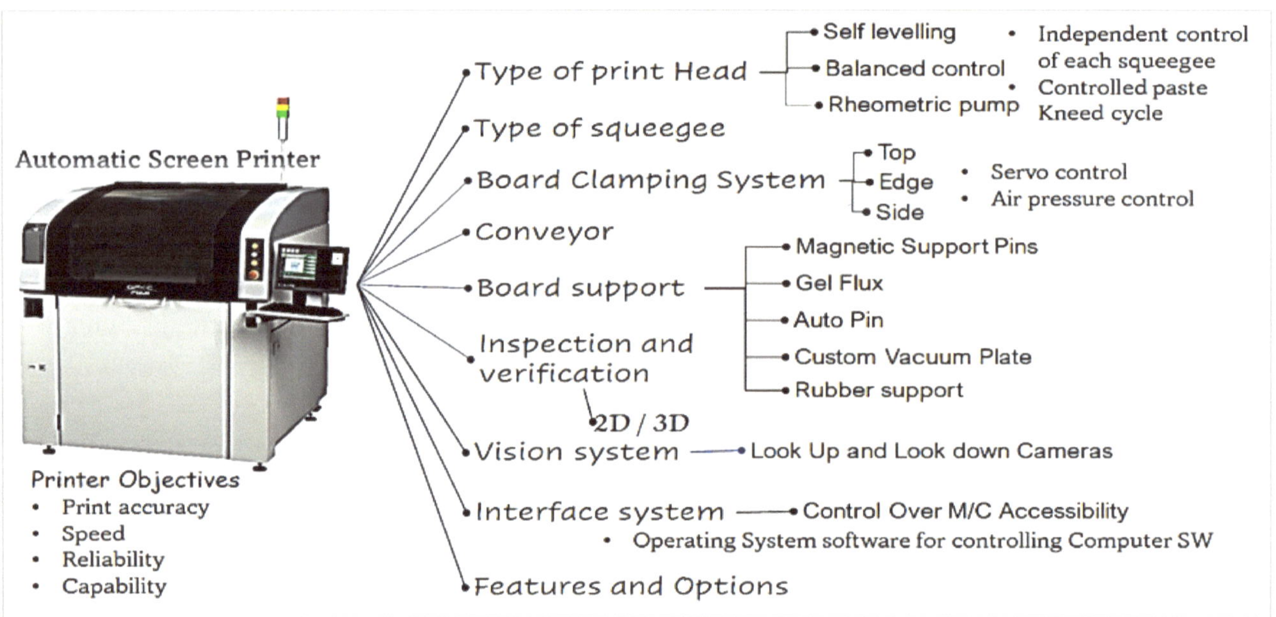

The electronic manufacturing industry adopted three types of technologies for Squeegee printing heads: self-leveling, Balanced Control, and Rheumatic. The user's selection of technology depends on Cost, Volume, accuracy, and defined quality acceptance levels.

Screen Printer –Print head types

Self Levelling Print Head

Balanced control print Head

Closed Loop pressure control system

Self-leveling Print Head

Most preferred head type in the printer by Electronic PCB assembly manufacturers. Center pivoting head. Squeegee pressure, print speed, and down stop are independently adjusted for the front and rear squeegee. Pressure and down stops are driven by a precision stepper motor. Easy maintenance and user-friendly.

Balance Control Print Head

It is a frictionless electro-pneumatic control to ensure the most accurate and repeatable pressure control for precise paste deposition. Designed with Automatic Squeegee down stop. Actual print pressures are monitored (E.g., 5 times/Sec) during the print stroke and are adjusted in real-time using closed-loop feedback control. Print speed, Blade attack angle, squeeze down, and Left and right print pressures were independently adjusted in both squeegee blades.

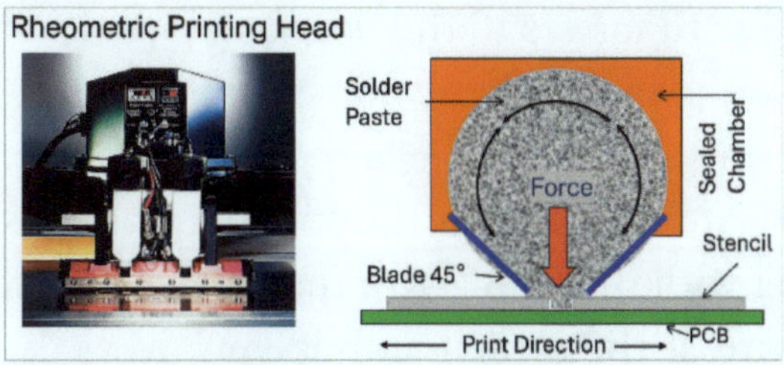

Rheumatic Pump Print Head

Rheumatic is a quantum leap in the printing process. Provides the pressure control of paste application. Delivers a whole new level of process control. Consistent paste dynamics are applied for each print because the paste is no longer exposed to air. 98% Paste Saving Over Standard squeegee blades. Paste rolls in a closed chamber and contains an inside sealed head, which ultimately contacts the stencil at the point of application. The paste is pumped into the stencil aperture for the circuit board pad when shared from the main body of the paste by the leading edge of the squeegee blade. Multiple head sizes for different printable board sizes. (E.g., 8",10",12",14", 16", and even up to 18").

CONVEYOR SYSTEM

Printers available with single conveyor or multiple conveyor systems for PCB transport, adapted with panel clamping, different clamping systems either Top or Edge. Top clamping is the most common. Either servo control or Air pressure control is used to clamp the Panels.

Board Support Unit

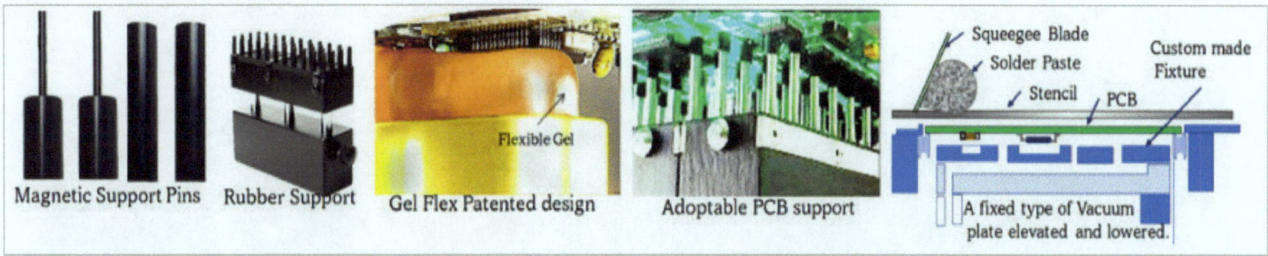

This is an important factor in ensuring the PCB is held flat against the stencil during the printing process. If not flat, printing defects arise – poor paste deposits and smudging. Magnetic support pins are generally supplied with a printing machine as a standard practice, which has a fixed height to have programmable positions and to ensure a constant process. PCB Blocks, also called rubber support, are designed as per machine specs, normally 30mm wide and magnetic. The position of the blocks can be changed depending on the size of the PCB.

Using Adoptable PCB Support is another way of supporting PCB Boards, available in different designs. It consists of supporting pins and the movement of pins based on spring action, and molds themselves with the components already soldered, the bottom side of the PCBs. It means self-aligned with the height of the components. In some variants of this type, individuals' pins are programmable for UP and Down positions.

PRINTER – VISION SYSTEM

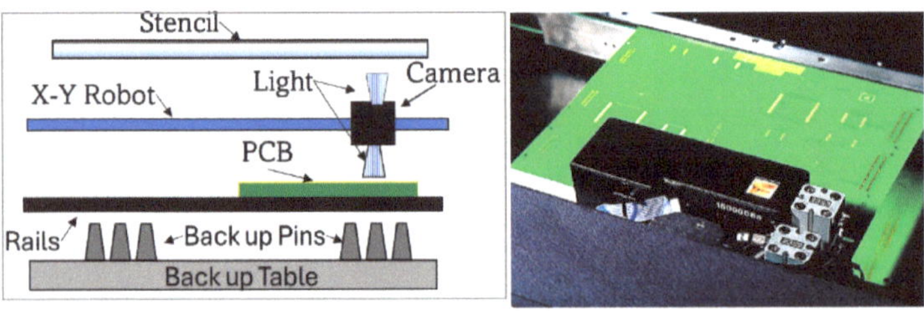

The vision system uses a camera, which moves over the entire printing area between the board and the stencil. The board moves to the board stop position either with a hard stop or with an ultra-sensor and then clamped. The backup plate, along with the tooling pins, was raised to PCB height and then raised to vision height along with the PCB. With the table at vision height, the camera lighting system allows it to look upward at the stencil and downward at the board. The moving camera locates the fiducials on the board and compares them with the fiducials on the stencil. For any offset within the limits, the print table aligns the board in the X, Y direction and Theta angle as per measured offsets to align the board precisely under the Stencil. Once aligned, the worktable further moved upwards to a print height where the PCB was attached to a stencil with zero gaps.

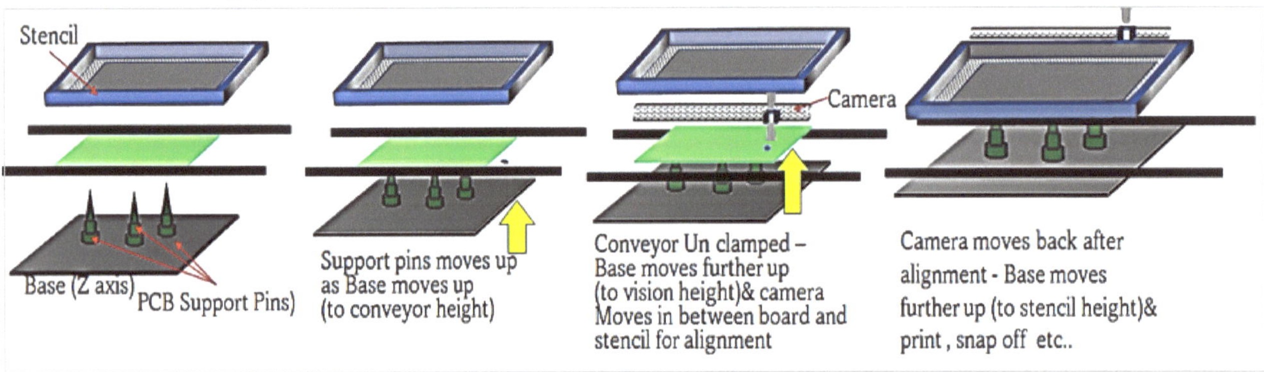

Image1 – Bare PCB attached to Stencil (Top view)

Image 2 – After Paste application by squeegee (Top view)

Image 3 – PCB releasing from the stencil Snap Off (Top view)

PRINTER – SQUEEGEE CALIBRATION

The printer uses one of two methods for applying squeegee pressure: Open loop or closed Loop

Open-Loop Control

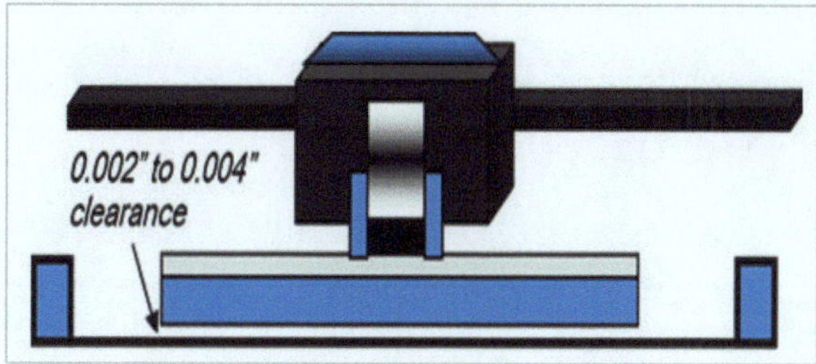

With this option, the system applies pressure calculated from stored parameters. To calibrate squeegees in the open loop method, as a thumb rule, the operator puts a piece of paper (paper thickness approximately 0.004 inches) on the stencil and initializes squeegees to begin the calibration process.

By using the Course and Fine incremental moves on the operational panel, move the squeegee down to the stencil till it just touches the stencil surface. To make sure of clearance between the squeegee and stencil, move the squeegee down till the paper just slides between the stencil and squeegee. At this point, the clearance normally is between 0.002" to 0.004". At this point, the measured values are stored in the system.

Closed Loop Control

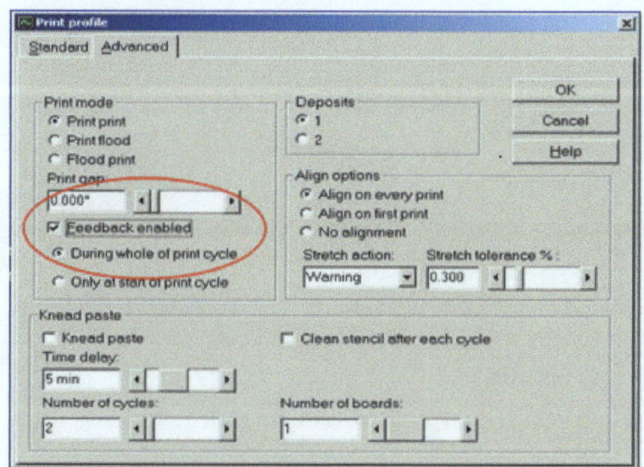

The pressure is constantly measured and may be adjusted during the print stroke. In this case, you need to calibrate zero pressure for both front and rear squeegees. This calibration applies to a pair of squeegees and is stored in the board file. You can remove and refit the same pair of squeegees without the need to re-calibrate. Feedback enabled will impact cycle time. Feedback during the whole print cycle will further slowdown the process.

Printer – Teach and Set Parameters

As mentioned earlier in this report, good print Quality in terms of accuracy, height, and paste coverage on a pad depends on the Equipment capability, features, and set parameters like print speed, print pressure, print stroke, and Snap Off. Here, we talk about some guidelines for teaching functions in stencil parameters. The teach function is also a multipart approach to set up a board profile. This approach requires you to teach the system before setting the print set parameters.

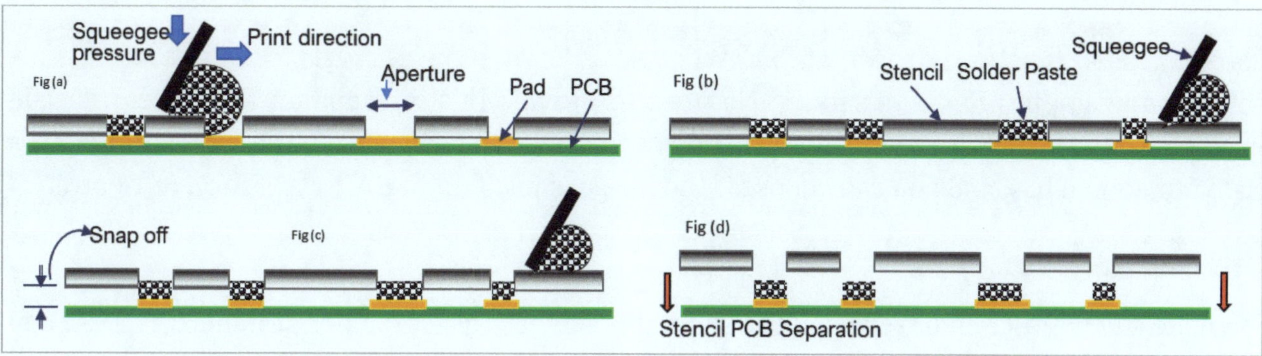

Teach Parameters

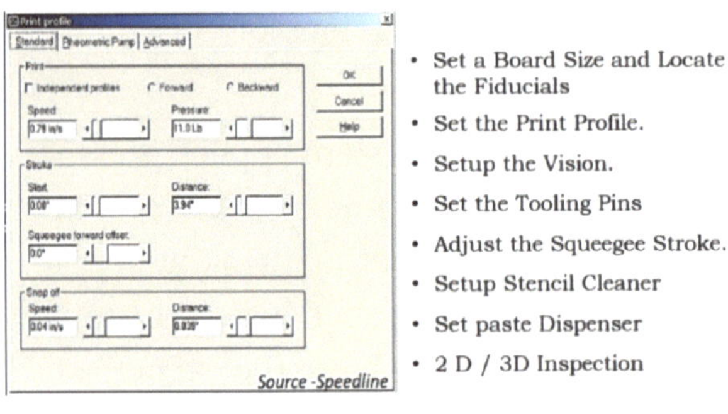

Source -Speedline

- Set a Board Size and Locate the Fiducials
- Set the Print Profile.
- Setup the Vision.
- Set the Tooling Pins
- Adjust the Squeegee Stroke.
- Setup Stencil Cleaner
- Set paste Dispenser
- 2 D / 3D Inspection

Print speed - Determines how quickly the squeegee moves across the stencil. The solder pastes need time to roll the aperture on the stencil.

Print Pressure/Squeegee Pressure - Force applied to squeegees during printing stroke.

Snap Off - After printing, the table moves down a few millimeters to separate the board from the stencil.

Snap–off speed - The speed at which the table moves down.

Snap–off distance - The distance over which the snap-off speed is maintained.

Print stroke - Start is the distance from the fixed rail to the position of the squeegee where the stroke begins or ends.

Distance - The distance that the squeegee travels when printing a PCB board

Setup PCB Information

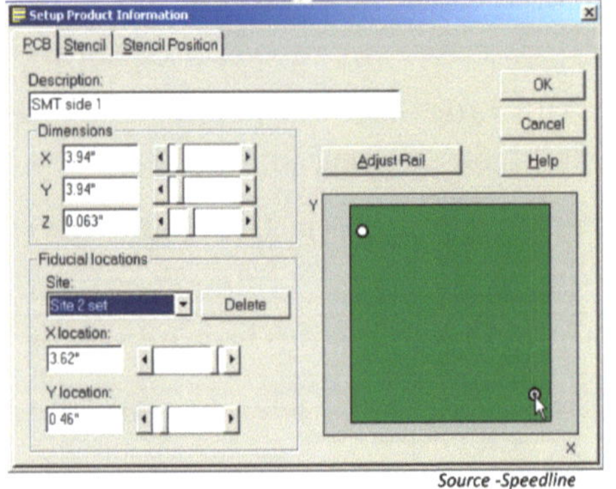

Source -Speedline

Provides controls to enter the Board description and dimensions and coordinates Board fiducials

X- Length in contact with the conveyor belt. The X dimension is used to calculate the position of the Board stop.

Y- Width between the front and rear rails. The Y Dimension positions the conveyor system. By default, the rails are 0.5mm (0.195in) wider than the programmed board width.

Z- The Thickness or depth of the PCB. The Z dimension calculates the printing position of the table when it is raised to printing height.

Whereas Automatic Screen Printers require input control parameters, the PCB dimensions' length (L) for Auto Board stop, the Width(W) for Auto Rail width adjustment, and the Thickness(T) for the Auto clamping.

STENCIL INFORMATION - SETTINGS

This window applies to old machines. No need to provide this information for present-decade printers. The Stencil Tab allows you to set up your stencil. You can choose the number of images that your stencil has, and you can choose if the image is justified in the Rear, Front, or Centre.

Setup Fiducial Teaching

To create a new program or recipe, it is a general practice to teach the board fiducials and stencil fiducials. Best teaching and setting tolerances are the best guidelines for smooth production.

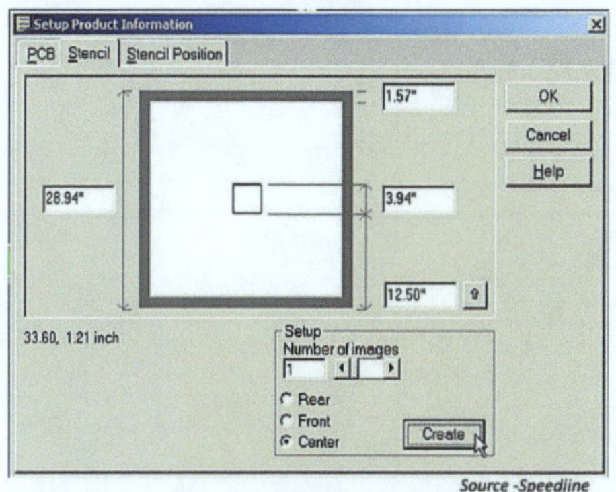

Source -Speedline

The machine prompts the user to load the PCB and stencil manually, and by using arrow buttons, moves the camera to locate the 1st fiducial on board in the center of the field of vision, and prompts the machine to learn it and subsequently learn the stencil 1st fiducial. The same procedure is to be repeated for the 2nd fiducial board and stencil. The machine prompts you to learn both the Board and the stencil Fiducials once the target matches the center of the Field of vision of the Machine vision system.

If the system sees more than one item in its field of view, it may not be able to distinguish between them when locating a Fiducial, you will need to learn to do it manually. In this case, the parameters "Threshold, Brightness, and Contrast" need to be adjusted to distinguish the Fiducial.

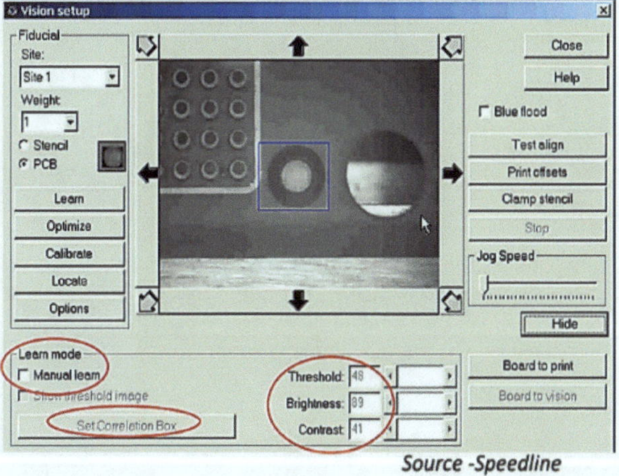

Source -Speedline

There are two basic methods that the vision system uses to locate fiducials. Boundary tracking mode and Correlation mode. In Boundary Tracking Mode, the system outlines the shape of the fiducials and stores the values for the area and perimeter of the fiducials. In Correlation Mode (normally the default mode of finding a Fiducial), the Correlation search mode uses a statistical method of comparison to find a *model* of the Fiducial image. When a Fiducial is learned, a box is drawn around its image. The entire contents of the box are stored as a Fiducial model. So, during *Locate*, the vision system moves the stored *model* around the field of view, searching for a match within the set tolerance.

STENCIL CLEANING

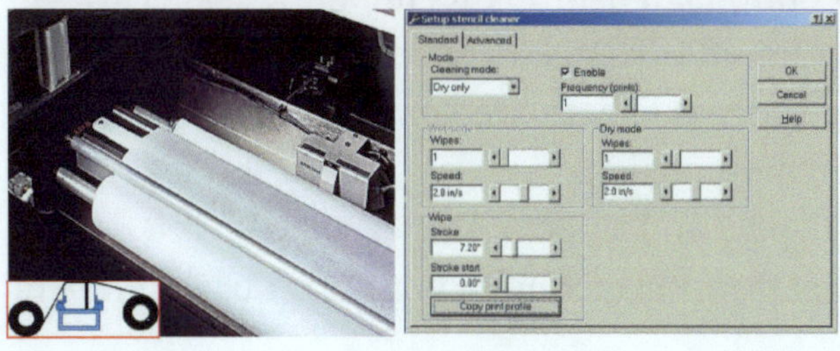

The equipment uses the cleaning fluid and roll paper to clean the bottom surface of the Stencil(mask). Dry /wet or Dry/ Vacuum can be selected as per the need, and the sequencing of wet and dry wipes. Equipment User can set the various control parameters such as **Wipe Stroke,** i.e., Distance from where wipe starting point, Duration of paper contact with the wetting bar before a wet cleaning cycle, i.e., quantity of fluid applied to the paper, and Control of the paper advance mechanism during stencil cleaning.

CONCLUSION

We need to change our perspective on Solder Paste printers. Equipment matters. Only appropriate equipment to be able to build quality boards with excellent print accuracy, particularly small pattern printing. Present decade, Original Equipment Manufacturers offering various options to cope up with speed, reliability, and capability. Features like automatic pressure adjustment, servo control, air pressure control panel clamping, and a Graphical user interface for easy understanding and traceability, i.e., information on squeegee solder and stencils. Even OEMs provide auto stencil adjustment and SPI Closed loop system in which Solder Paste Inspection results are analyzed and reflected in the printer.

Jet Printing technology is also one of the good options in printing technology, especially for Flex and Rigid-Flex assemblers. Very easy to program and changeovers very fast. No need for stencils, which in turn avoided the Stencil cleaning process. There are some concerns with this technology, such as equipment maintenance and higher consumables cost, because dot printers showed good performance for the solder pastes of Type 5 and Type 6. Hence, 70% of assemblers used stencil printing, 20% of assemblers considered jet printing technology, and 10% of assemblers used this technology.

FORWARD

Solder paste printing is the first step in PCB assembly manufacturing. Solder paste applied on the pads of the substrate / PCB for soldering surface mount devices by using the Screen Printer equipment. The equipment screen printer requires good design stencils and squeegees, which are vital for controlling the print process parameters. Stencils generally involve the use of a "trailing edge" with metal squeegee blades with 'on contact.

Often confused with the mesh print and stencil print, both processes require similar machine platforms with engineering controls and vision parameters. The mesh screen process is outdated and limited to some applications, which involve polyurethane squeegees with 'Off contact', i.e., a gap between the mesh and substrate.

Stencil printing became popular in the 1980s as aperture pattern technology improved to enhance the solder paste release with control specification limits concerning volume, height, and pad coverage.

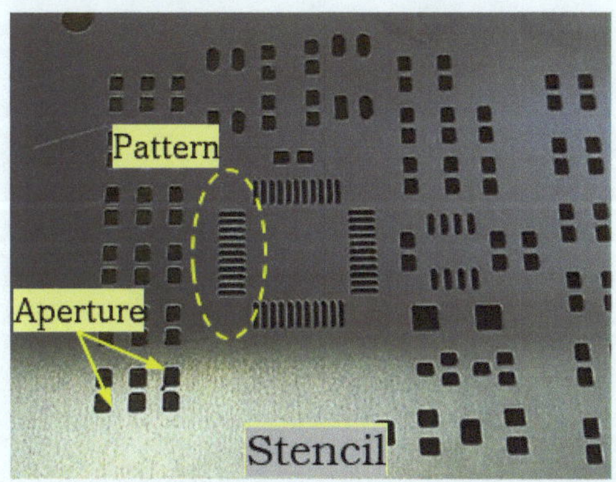

Solder Paste Printing in electronic assembly is the process of using stencils, often made of thin sheets of stainless steel or Nickel material with foil thickness ranging from 1 mil (0.001 inches) to 7 mils. Foil thickness of 8 mils and above becomes limited for special applications. Suitable thickness will be determined by the desired amount of solder paste and height. By using a stencil, Solder paste is applied on the PCB pads instantly, with the help of Squeegees.

The stencil contains a bunch of holes in it, called Apertures, which represent all SMD footprint pads of the board layout. It has a circuit pattern cut into it that matches the pattern of surface mount devices required on the PCBs so that the assembler can apply solder paste only where it needs to be.

ROLE OF STENCIL IN ELECTRONIC ASSEMBLY

As the size of the apertures becomes smaller and smaller, due to the need for fine pitch component placement, stencil design becomes quite complex and must be adherent to IPC specification 7525 and other related standards.

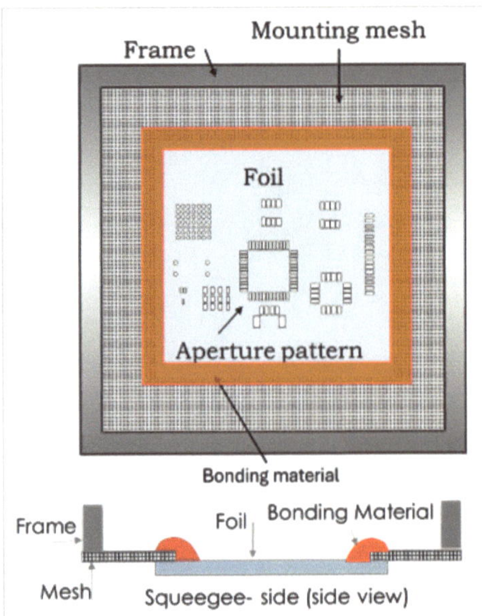

Framed stencils, also called "glue-in "or mounted stencils, are designed in such a way that Foil is securely mounted to either a cast or hold-out aluminum stencil frame using a mesh border to get high stability, high consistency, and high mesh tension. All standard stencils are manufactured with a foil made of stainless steel glued with high-precision mesh. The standard stainless-steel screen is delivered in 80 mesh (mesh/inch) and is tensioned at a 90° angle. The diameter of the screen wire is 0.1 mm. The SMD stencil is delivered with a tensile strength of 40 N / mm² to achieve optimum printing results.

Framed stencils are available with frame sizes (square) 23"x23", 25"x25", and 29 x x29 inches. As a standard practice, Screen printer equipment manufacturers, OEMs, design their equipment to hold a stencil size of 29"x29". Options available to hold/use 23" X23" or 25" X25" frame stencils with the help of suitable fixtures. Some screen printers use rectangular framed

stencils for the paste printing process, especially for LED PCB solder paste printing. Hence, choosing the shape and size of framed stencils purely depends on the paste printing equipment's capabilities. Present technology, Screen printing equipment has the facility of Auto Frame Adjustment to handle all frame sizes without the requirement of Fixtures / adaptors.

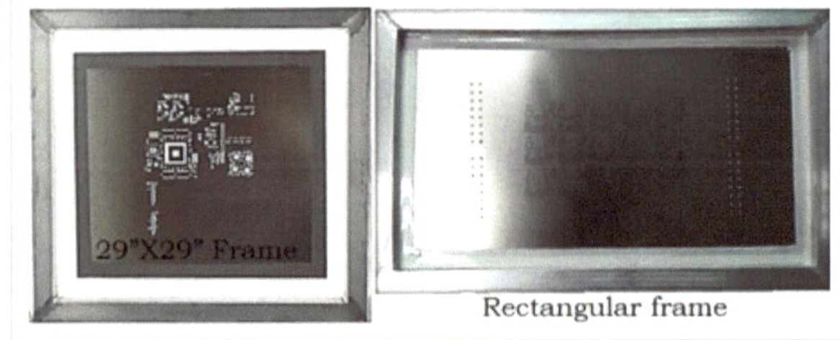

SMT manufacturing industries use many types of stencils for their intended applications such as stencil printing, rework stencils, adhesive stencils, step stencils, and 3D stencils.

Adhesive Stencils

The adhesive is applied to hold the SMT components to the bottom side of mixed technology boards during the wave soldering operation. Adhesive stencil printing is one of the options widely accepted process instead of high-speed dispensing machines. Stencils with a thickness of 150 µm - 400 µm are preferable to use. The diameter, height, and number of adhesive spots are determined by component geometry, and the geometry data is provided from the pick and place process or an in-house library. Different types of adhesive stencils are designed as required – with round aperture, separated areas for the print process, for "wet" in "wet" printing, and for exact adhesive print heights.

Step-Up Step-Down Stencils

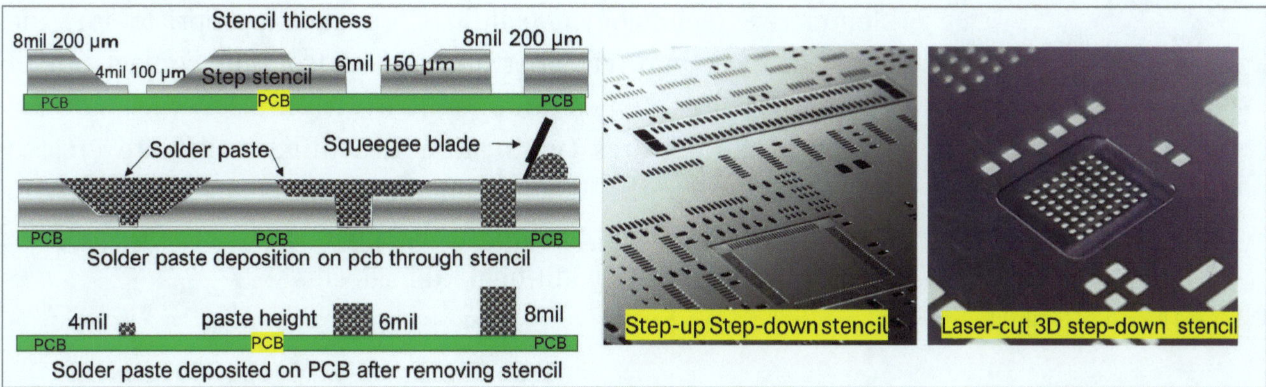

Step stencils are more than one foil thickness. In the case of step-down, material thickness is reduced in areas to lower the volume of paste deposited, and for step-up, material thickness is increased to increase the volume of paste deposited, such as BGA, edge connectors, D Pak, etc.

3D Stencils

This type of stencil is made with either Aluminum, Acetal, or Glass epoxy sheets. Used to print on PCB after through-hole assembly. Can achieve a height difference between 20µm – 3mm. Polyurethane squeegees are recommended for these kinds of Stencils. Used for printing on through hole and clinched boards.

Stencil Fabrication

The Holes or apertures of the stencil are created by any of the following viable methods

1. Chemical etched technology

2. Laser Cut Technology

3. Electroforming.

Each method has advantages and drawbacks in terms of Cost, accuracy level, transfer efficiency, and stencil life. Based on the required specifications, any of the above technologies can be opted for their application.

Chemical itch

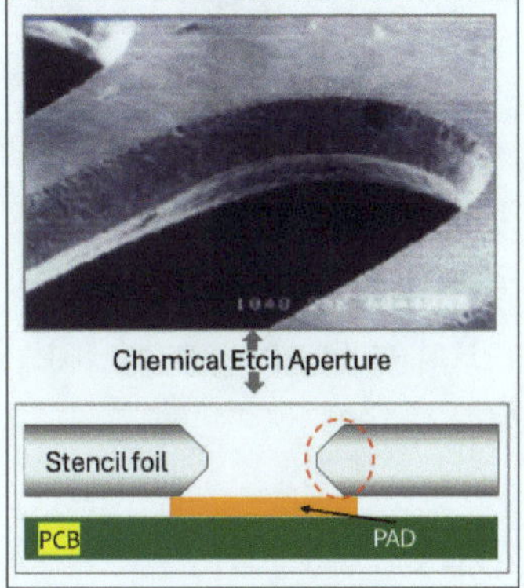

Cleaned metal surface chemically etched from both sides of foil center simultaneously. The aperture etch form from the image. Aperture walls are usually hourglass-shaped. Fabrication cost is cheap, and this type of stencil is appropriate for pitch sizes > 0.63mm. This can withstand the known standard cleaning method. Due to the chemical etch process, aperture side walls are rough in nature and hence not recommended for fine pitch due to poor release of paste. Stencil rework cannot be possible for this type.

Laser cut stencils

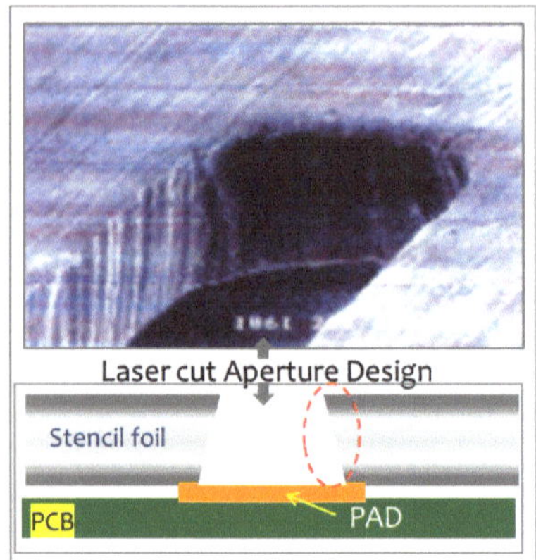

Subtractive technology, Medium Cost, and appropriate for fine pitch size > 0.4mm. The apertures are cut from the contact side of the stencil, then flipped and mounted with the squeegee side up. Trapezoidal apertures (with more structured walls) are created automatically by the laser beam focus, which in turn helps to enhance solder paste release. This is only one process that permits rework, enlargement, additional fiducial, etc.

Electroforming Stencils

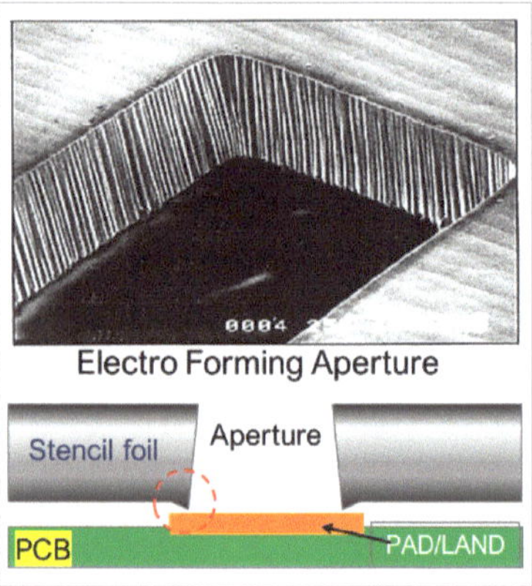

Electroforming is a process of producing metal/stencil/parts by depositing atom by atom. Precise and controlled process. Highly accurate and aperture tolerance ±0.0003". Smooth mirror-like Tapered aperture walls, provide less surface area (transfer efficiency > 95%) for solder paste to cling to. The special Gasketing feature results in less needed wiping, Nickel electroform offers a lower coefficient of friction compared to stainless steel. Used for fine pitch components (20 mil to 12 mil pitch). Also used for μBGA s, Flip Chips, and Wafer Bumping (12 mil to 6 mil pitch).

STENCIL SOLDER PASTE TRANSFER EFFICIENCY

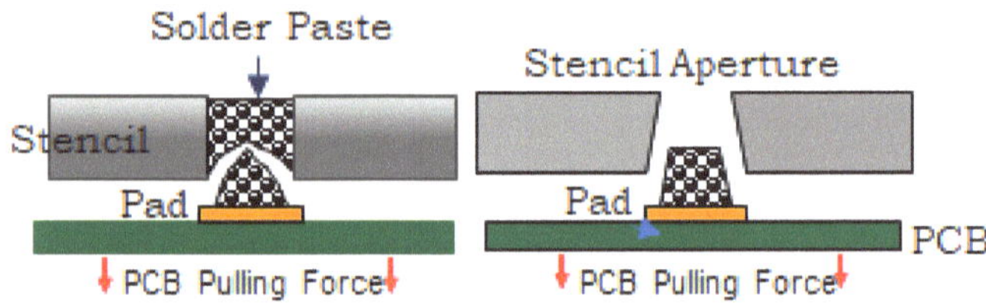

Transfer efficiency is a measure of how much of the solder paste that fills the stencil aperture will transfer or print to the surface of the printed circuit board. Stencil design is the critical factor in determining the transfer efficiency of a solder paste. Smaller stencil apertures require the highest possible solder paste transfer efficiency. The holding force of the solder paste to the printed circuit board pad must be greater than the holding force of the solder paste to the stencil aperture walls for the solder paste to release from the aperture and attach to the printed circuit board pad.

Stencil Design Guidelines

The size and shape of stencil apertures determine the volume, uniformity, and definition of the material deposited onto substrates. Measures such as Area Ratio (the area under the aperture opening divided by the surface area of the aperture wall) and Aspect Ratio (aperture width divided by stencil thickness) can be used to determine appropriate aperture sizes.

$$\frac{Pad\ Pulling\ tension\ (P)}{Relating\ Wall\ Tension\ (R)} = \frac{Aperture[Length(L)\ X\ Width(W)]}{Stencil\ thickness\ (T)\ X\ Aperture\ Perimeter\ \{2\ X\ (L+W)\}} = \geq 0.6$$

The general rule is that, for acceptable paste release, the area ratio should be greater than 0.66 and the aspect ratio greater than 1.5. When designing apertures that adhere to these rules.

When the pad area is greater than 66 percent of the aperture wall surface area, the probability of achieving efficient paste transfer is increased. As the ratio decreases below 66 percent, paste transfer efficiency decreases and print quality becomes erratic. The finish of the aperture walls can have an impact at these levels.

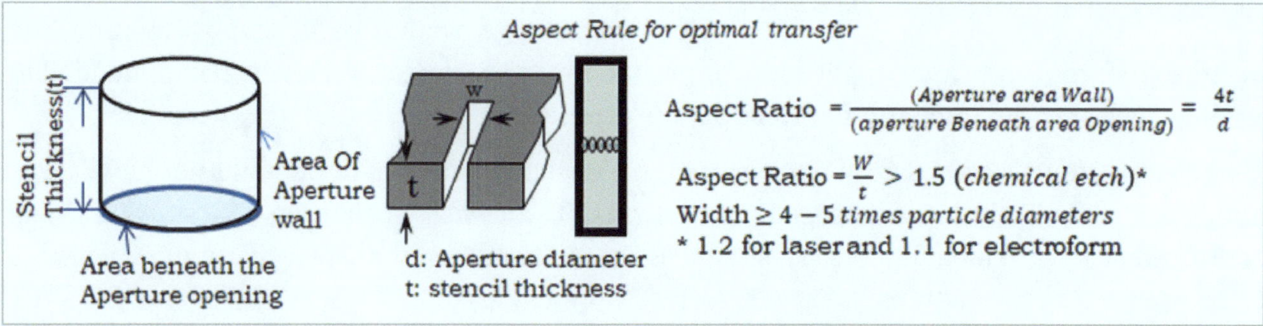

Area Ratio - Ensures that the forces pulling a material onto the pad are greater than the forces holding it in the aperture.

Stencil– Comparing Circular and Rectangular Apertures

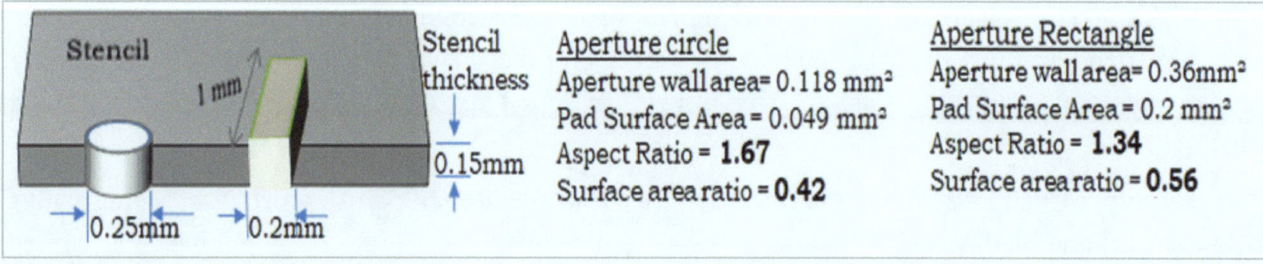

The circle is more difficult to print because of the lower Surface Area Ratio. The aspect ratio is no longer a good indicator as here it suggests that the circle would be easier to print - isn't it? Design Stencils for Lead-Free Paste have higher surface tension and do not Wet or spread on the surface of Pads as easily as Eutectic Solder. Modify surface aperture design to increase pad coverage on pads.

PITCH mils(inch)	PAD WIDTH	APERTURE	STENCIL THICKNESS	ASPECT RATIO
25(0.635)	15(0.38)	12(0.3)	6 (0.15)	2.0
20(0.5)	12(0.3)	9 – 10 (0.23 – 0.25	5 – 6 (0.13 – 0.15)	1.7
16(0.4)	10(0.25)	7 – 8 (0.18 – 0.20)	5 (0.127)	1.4
12(0.3)	8(0.20)	5 – 6 (0.13 – 0.15)	4 – 5 (0.1 – 0.13)	1.2

BGA	Pad	Aperture	Thickness	Area Ratio
60 mil	32	30	6 - 8	1.25 - 0.94
50 mil	25	22	6 - 8	0.92 - 0.69
20 mil	12	10	5 - 6	0.50 - 0.42

STENCIL HANDLING AND STORAGE

Some general guidelines for stencil handling are mentioned below. They are particularly useful for improved life and better performance of stencils. For stencils having Fine pitch apertures (0.3mm) and micro-BGA, manual cleaning is not recommended. Need to remove solder paste trapped within the apertures before storing after application, or production, or changeover. Hence, recommended to use Stencil cleaner equipment for perfect cleaning results.

Cleaned stencils after use, must be stored in a designated area. They should not be left out, as they are more likely to be damaged. Inspect stencils for wear or damage before using them. Identify stencils with job numbers. This reduces the mishandling or misplacement of stencils.

Squeegees

It is well known that stencil printing is a complex process influenced by several variables that include hardware, software, materials, and process-related factors. Squeegees play a vital role in the printing process to maintain the accuracy of paste deposition of solder paste volume for the components 01005 passives, CSP/ BGA, and do the complex wafer bumping processes.

With squeegee blade printing, only two print process parameters can typically be controlled: squeegee speed and downward squeegee pressure. There are two primary methods of applying solder paste to a circuit board using a stencil printer: squeegee blade printing and enclosed head printing.

Two types of squeegee blade materials are used in the stencil printing process: Metal blades and Polyurethane blades.

Polyurethane Squeegee Blades

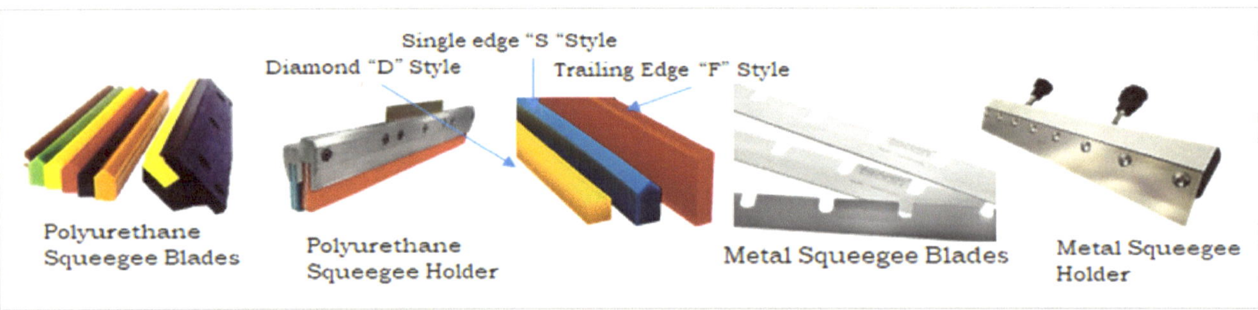

Polyurethane blades with a high durometer rating (90-110) have shown success in many applications. The top edge must be kept sharp all the time. If it wears, the squeegee pressure needs to be increased to prevent the mapping of the paste.

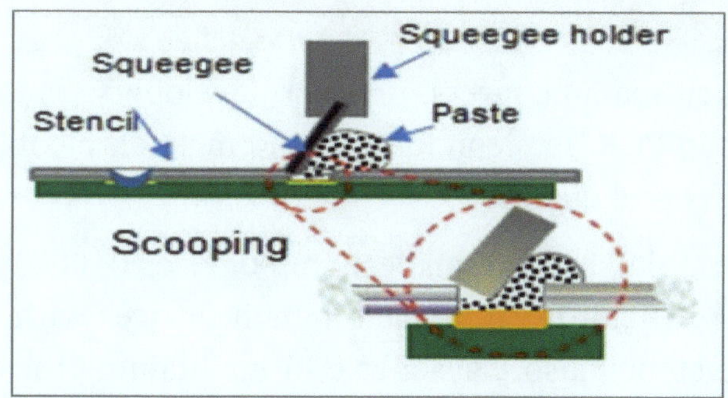

Metal Squeegee Blades

In general, only two metal blade printing parameters that affect aperture filling can be controlled: squeegee speed and downward squeegee pressure. The speed should not be set so high that the paste does not roll as it moves across the stencil. The blade pressure is usually set so that no paste remains on the stencil behind the squeegee. Lasts longer than Poly Blades. Reduces the effect of Scooping or Scavenging.

Squeegee Length

The squeegee blade length should extend .5 to 1.5 inches beyond each end of the PCB. (D should be > 0.5 inch and < 1.5 inch)

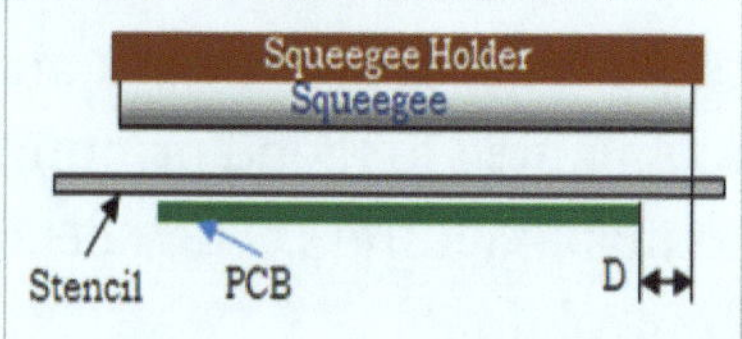

Squeegee Attack Angle

The angle between the blade and stencil before the print stroke is the contact angle; During the print process (with print pressure and speed in active mode) the angle between the blade and stencil is known as the attack angle.

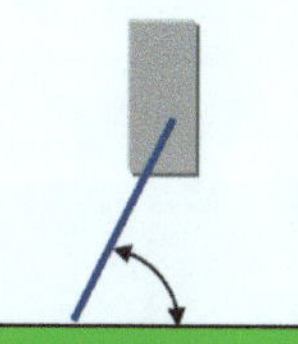

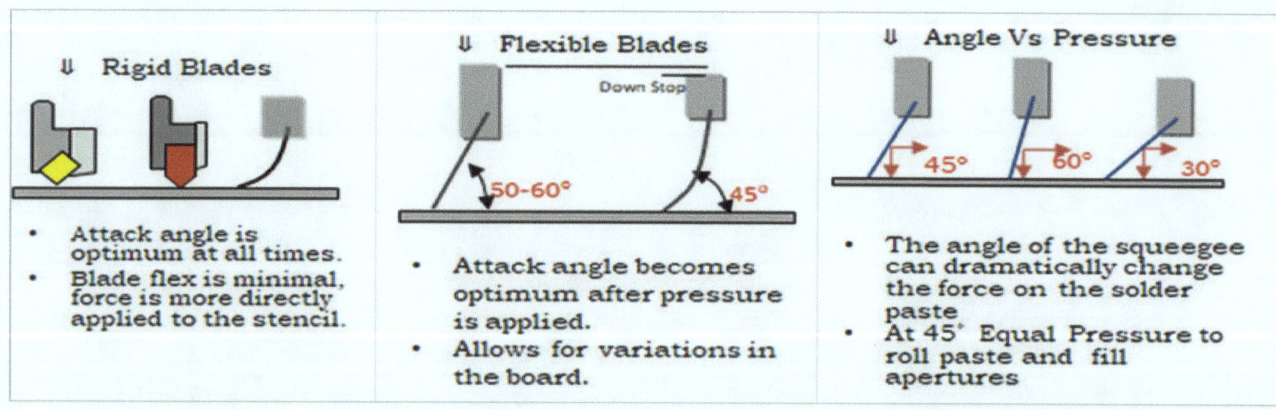

Paste Rolling Effect

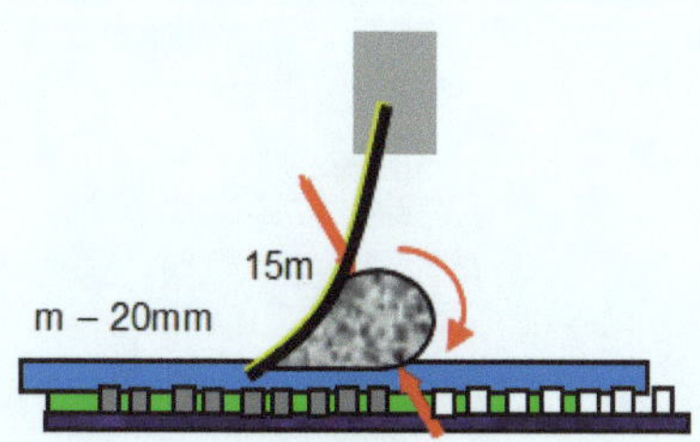

To achieve a better paste rolling effect, you need to maintain a paste roll diameter of 15mm to 20mm (thumb of a rule). The frequency of paste addition is to be maintained as per the paste consumption. Appropriate squeegee length also maintains the paste-rolling effect. Proper setup of Squeegee side dams (side supports) reduces the paste wastage on stencil while printing. Lift Height of 1" can help to minimize paste sticking to the blade.

SUMMARY

Paste transfer efficiency is a major concern for stencil manufacturers, product developers, and designers. Reason – miniaturization. Miniaturization in PCB, semiconductors, and most efficient product design without compromising on functionality and quality.

Stencil aperture design is becoming the most complex for the smallest chip devices 01005" and 008004". Paste Volume deposition is even more difficult for Integrated circuit devices with less than 0.3mm pitch. The further thickness of the stencil also plays a role in maintaining the height of the paste deposit on the pad.

Somehow laser cut, electroforming stencils are popular in meeting the demands. The type of paste, squeegee attack angle, print pressure, print speed, and snap-off parameters need to be maintained for good printing results. Hence, tolerances for printing parameters cannot be granted to manufacturers. They need to adopt the concepts of statistical process controls to maintain the control limits.

FORWARD

Stencil printing is a process of depositing solder paste on the printed circuit board to establish an electrical connection. The equipment and materials used in this stage are a stencil, a Squeegee, **solder paste,** and a printer. Stencils are used in SMT for paste print applications. Stencils are unique to the product model and the PCB Model. The selection of stencil depends on thickness and the Aperture design. Further details about the stencils and squeegees are explained in the following section.

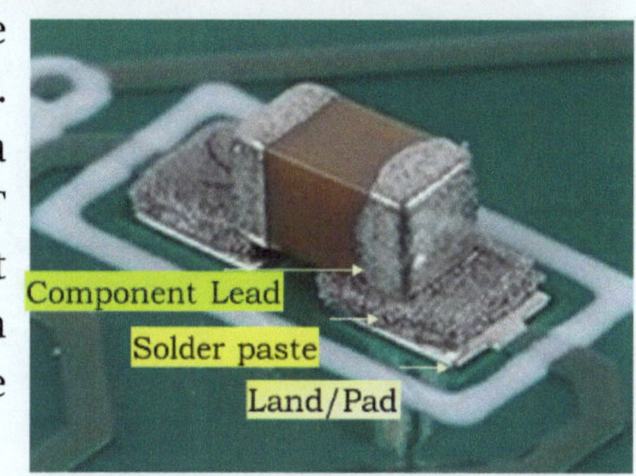

SOLDER PASTE

Solder paste is a consumable material used in printed circuit board assembly to connect surface mount devices to pads on the board. The best suitable Solder paste depends on the key requirement of the Manufacturer. Solder paste selection depends on application, cost, and process requirements. The process differs for Tin-lead solder paste and lead-free solder paste. Due to rising concepts of "Go green", manufacturers face constant demand to adopt alternatives to environmentally friendly manufacturing processes, especially for using RoHS (Restriction of Hazardous Substances) compliant components and Lead-free Solder usage. It is a proven fact that most of the defects (Approx 70%) in circuit-board assembly are caused by issues in the solder-paste printing process or due to defects in the solder paste.

Solder Paste is a mixture of Powder form + Resins (Rosins) + Activators + Solvents + thickening agents + Antioxidants and rheology aids. Upon heating, Solvents evaporate, and activators attack the metal surface, resulting in Cleaning. Further heating – solder powder particles melt and form a liquid mass, which makes the solder joint. The composition of the above mixture depends upon its intended use.

Solder Paste Composition

A solder paste is essentially a powder metal solder mixed in a thick medium called Flux. Flux is added to act as a temporary adhesive, holding the components until the soldering process melts the solder and makes a stronger physical connection. Solder paste is thixotropic, meaning that its viscosity changes over time with applied shear force (e.g., stirring). The *thixotropic index* is a measure of the viscosity of the solder paste at rest compared to the "worked" paste. Solder Paste is a composition of Alloy powder (90% by weight) and an organic chemical, also called Flux, cream – 10% by weight.

We should remember some known facts about metals before understanding the solder paste alloy composition and rheology of the Paste. Metals Oxidases with Oxygen, Nitrogen, Water (moisture) pollutants (Oxides of Sulphur in Air). Copper surfaces form Copper Oxides. Copper oxides interact with carbon dioxide and Moisture in the air to form Carbonates or hydroxyl carbonates. Nickel produces a continuous thin film of Oxide. Silver reacts with Traces of Hydrogen Sulphide to form silver Sulphide. Once the layer of these compounds forms on the surface of a metal, it causes the Passivity of a metal surface.

So, printed circuit board assembly (PCBA) needs printed circuit boards that have metal pads, Lands, and tracks. For the Solder joint, we use paste, which is a composition of alloys, which are all metals. Further, oxidation happens at a faster rate upon heating. Hence, all mandatory processes are simply against nature. But, without all this process, a solder joint cannot happen. So, the only option is to determine how we can control oxidation. Hence, control parameters can be PCB surface finish, solder alloy composition, flux, print process, and reflow process considerations.

The solder alloy must have nonhazardous properties, good wetting, be mechanically reliable, be thermally resistant, have a relatively low melting temperature, and be compatible with component lead surface coatings.

Both Lead-Free (SAC)and Leaded (Sn Pb) are categorized under Eutectic. usage of lead is almost eliminated to thrive the Go Green concept. Eutectic describes a solder alloy that melts and freezes at the same temperature. For example, SAC alloy shows Eutectic behavior at 217°C. Similarly, the Sn Pb alloy melts and freezes at 183°C. This is the reason these two solder alloys have their place in the Electronic Industry.

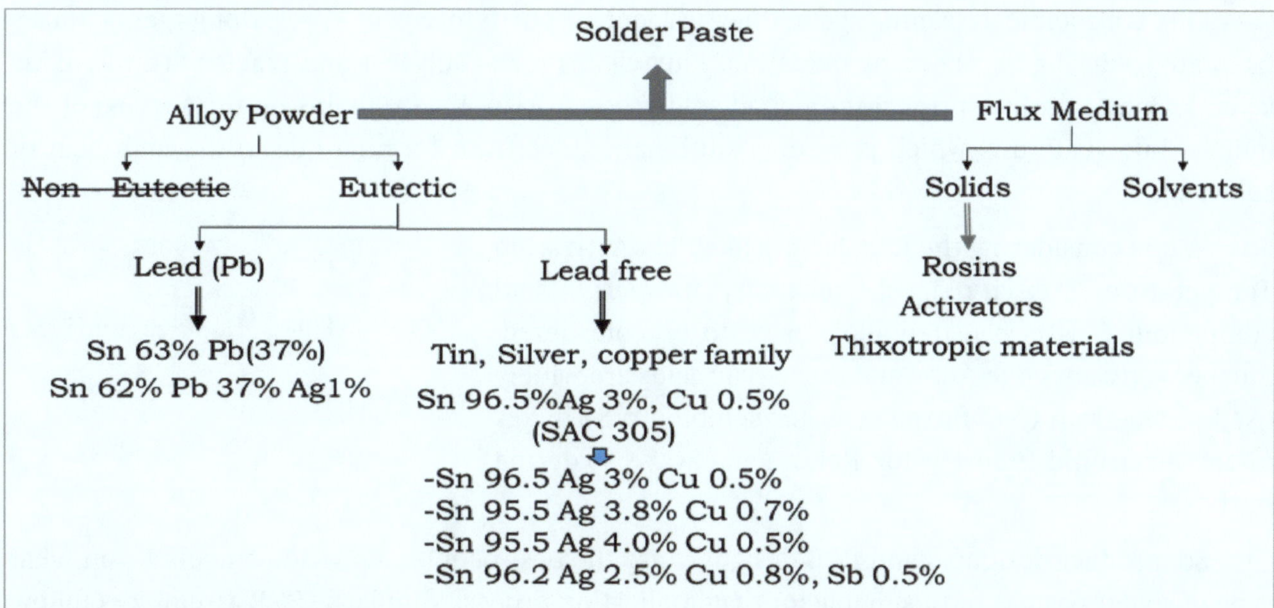

Non-eutectic describes a solder alloy that solidifies and melts in a different temperature range. In this range, the alloy is pasty with a Low melting temperature, hence not recommended for the electronic Industry. The solder alloy Sn Ag Cu family is most used in the electronic manufacturing industry because it matches the required solder qualities. A combination of the Tin, Silver, and Copper families is well accepted by assemblers as a good solder joint happening at relatively low melting points with good wetting and good reliability characteristics. Alloy element tolerances are maintained as per standard IPC -J-STD-006.

So, Solder paste is composed of a solder powder and solder flux. The resultant properties of the solder paste are directly established by the raw material utilized. The final property of the solder paste is controlled by raw material characteristics. Control parameters for solder powder are Size, shape, and Surface chemistry. The flux system is the control parameter for chemical purity, consistency, and viscosity.

FLUX

Metal surfaces need cleaning. Because they undergo oxidation on exposure to the environment and form compounds with Oxygen, Nitrogen, Water, and pollutants. In our scenario, Pads on PCB, Component leads are metal surfaces which are prone to oxidation. At the same time, we apply heat for sufficient temperatures to make a solder intermetallic bond. Which in turn furthers fast oxidation. Hence the requirement of flux – cream-like texture – is mandatory to clean oxides and metal surfaces.

Flux is a chemical cleaning agent, flowing agent, or purifying agent. As cleaning agents, fluxes facilitate soldering by removing oxidation. Flux cleans metal surfaces and reacts with the oxide layer, leaving a surface primed for a good solder bond. Also, Flux remains on the surface of the metal while soldering, which prevents additional oxides from forming due to the high heat of soldering.

While considering the flux, factors like Flux Activation Temperature, Activity level, reliability property, and compatibility with lead-free alloys need to be considered. Further activators like Alkyl and carboxylic acids are widely used in No-clean (NC)fluxes or water-soluble (WS) fluxes in the electronic Industry for Reflow and wave soldering purposes.

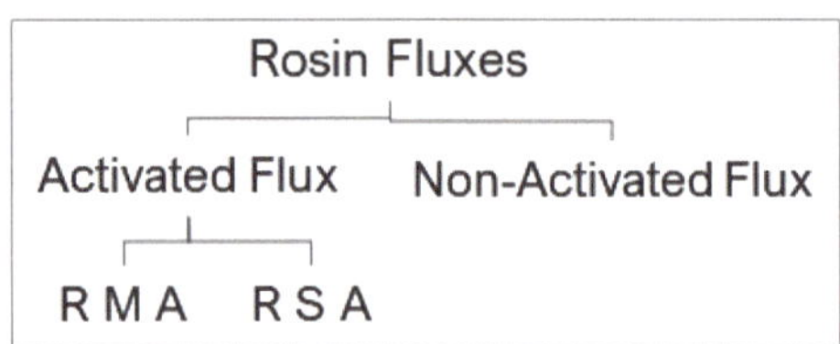

So, product designers and PCB designers are the best people to provide guidelines on what type of solder paste is most suitable for that application process. Similarly, PCB assemblers follow paste manufacturers' guidelines about Solder Paste Handling, print process considerations, and reflow process (profile)guidelines.

FLUX PROPERTIES

Rosin-based Fluxes are widely used, especially when it comes to the usage of the Wave soldering process, as they are natural organic compounds. It is a solid, sometimes vitreous, crumbly, with a light Yellow to Brown color. Rosin-based Fluxes is a complex mixture of organic compounds, mostly Terpenoids and Hydrocarbons. The most important compound is Abietic (85%)or Sylvic acid (C20H3002) + 'd' and 'l' – Pimaric acids (12%) + others (2%) Why Rosin? Because Rosin promotes wetting. Does not attack copper. Attacks only the passive layer on copper. Good vehicle for more active fluxing compounds (Amine Hydro bromides) and has good tact and rheological performance. More reactive towards the passivation layer of metals when adding activators such as Bromine Salts of Amino compounds and Carboxylic acids (activating agents) to rosin.

RHEOLOGY OF SOLDER PASTE

Ketchup not flow even when the bottle held upside down

Ketchup temporarily liquidifies and flows when struck or vibrated

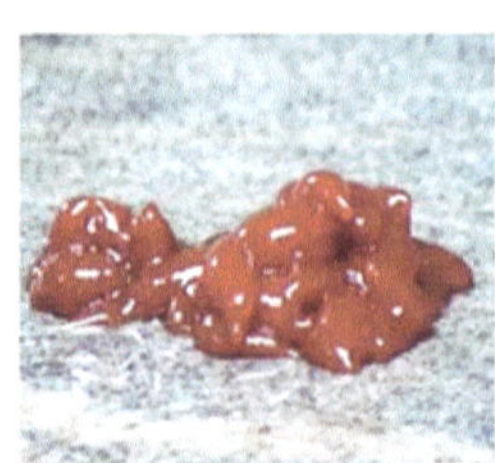

Once vibration stops ketchup returns to its original viscous state

It is a Science dealing with material flow properties under stress. It refers to the change in viscosity of solder paste as the shear stress is applied. Solder paste Rheology is influenced by the metal Powder content, particle size, particle Shape, Temperature, and Humidity.

VISCOSITY

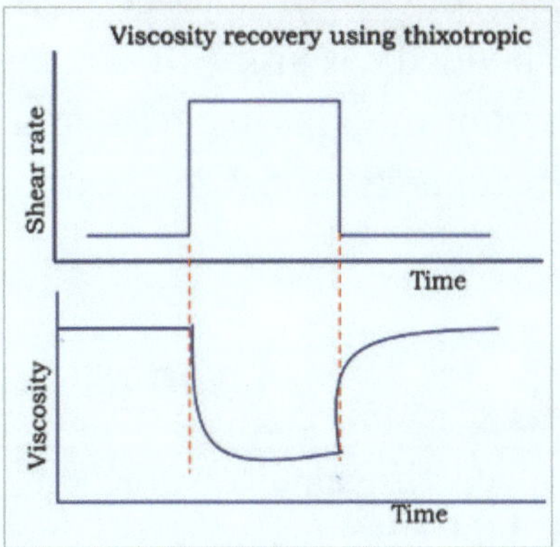

Viscosity is the degree to which material resists the tendency to flow. In this case, varying viscosities of solder paste are desired at different levels of shearing force. Such a material is called **thixotropic.**

"Thixotropic" is how solder paste changes viscosity with shear stress. When solder paste is moved by the squeegee on the stencil, the physical stress applied to the paste causes the viscosity to break down, thinning the paste and helping it flow easily through the apertures on the stencil. When the stress on the paste is removed, it regains its shape, preventing it from flowing on the circuit board.

SLUMP

Slump is the characteristic of a material's tendency to spread after application. That means deformation of the paste deposit after printing. Height reduces as the surface expands. Theoretically, the paste's sidewalls are perfectly straight after the paste is deposited on the circuit board, and it will remain like that until the part placement. A paste's slump should be minimized, as the slump creates the risk of forming solder bridges between two adjacent lands, creating a short circuit.

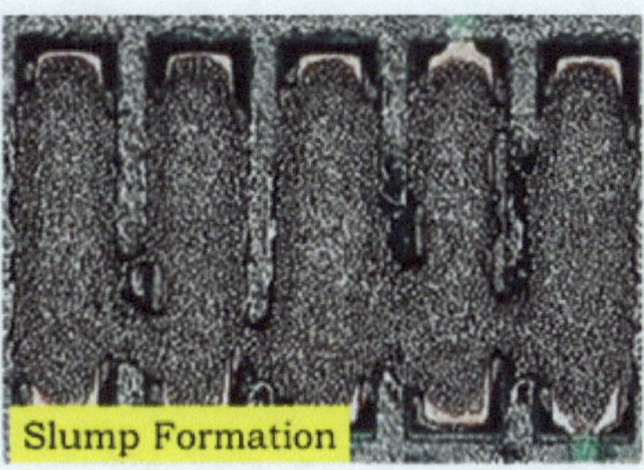

Slump formation also depends on the working life of the paste and lower viscosity. The working life of the paste is the amount of time the solder paste can stay on the stencil without affecting its printing properties. The paste manufacturer provides this value. Lower viscosity means solder paste creamer and less viscous causes defects such as solder beading, bridges, and slump.

SOLDER PASTE VISCOSITY

APPLICATION METHOD	VISCOSITY
Syringe Dispensing	200 – 400 kcps
Mesh screen printing	400 – 600 kcps
Stencil Screen printing	800 – 1200 kcps

Viscosity is a unit that measures a fluid's resistance to flow or an inverse measure of its fluidity. Viscosity is measured in units Centipoise(cps)or kilo centipoise (kcps). Recommended solder paste viscosity for solder paste supplied in Jars should be between 500kcps – 1200kcps (500,000cps – 1,200,000 cps) before printing. Hence, the recommended viscosity of solder paste for PCB assembly applications depends on the paste composition, the raw material, and the supply type. The flux system is the control parameter for chemical purity, consistency, and viscosity.

SOLDER ALLOY POWDER CLASSIFICATION

Solder Alloy powder is manufactured in various methods such as Gas, centrifugal, ultrasonic, etc. For example, in the centrifugal process, the Solder is melted in the solder pot situated at the top of the tank, and the molten solder is dripped directly onto a high-speed spinning spindle. When solder drops hit the spindle, it splash toward the wall of the tank, and before it reach the wall, the solder becomes spherical and solid. Chamber Purged with nitrogen, resulting in a very low oxygen density. Solder powder obtained at this production stage ranges from 1~ 100µm. After this stage, the solder powder is brought to the classification stages

Solder paste alloy powder particle size matters a lot in the selection of solder paste for different applications and soldering processes. Shelf life, stencil life, reflow performance, voiding behavior, reactivity, and stability are all affected by solder powder size. Even stencil aperture design guidelines are based on particle size.

Available in the market in different Types with a description: Type 1, 2, 3...Type 7 and classification based on particle size. Here are IPC guidelines for Size classification for the paste manufacturers and assemblers.

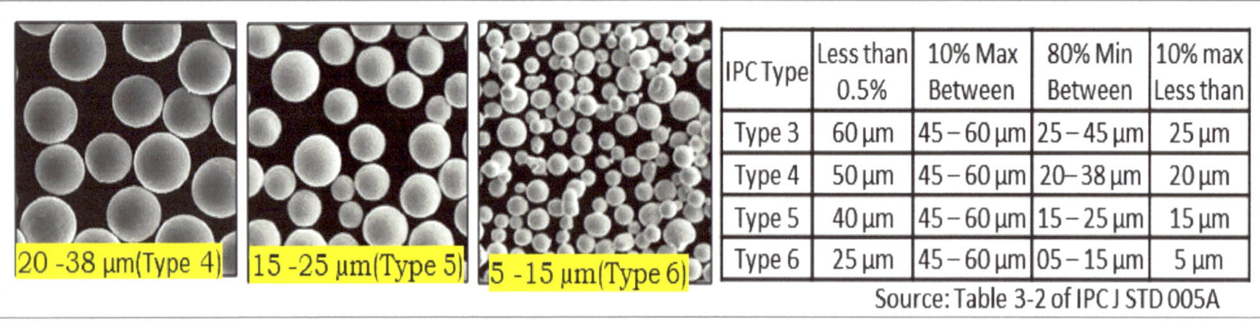

IPC Type	Less than 0.5%	10% Max Between	80% Min Between	10% max Less than
Type 3	60 µm	45 – 60 µm	25 – 45 µm	25 µm
Type 4	50 µm	45 – 60 µm	20 – 38 µm	20 µm
Type 5	40 µm	45 – 60 µm	15 – 25 µm	15 µm
Type 6	25 µm	45 – 60 µm	05 – 15 µm	5 µm

Source: Table 3-2 of IPC J STD 005A

Most assemblers use Type3 0402" printing, Type4 for applications 0201", µBGA, etc., and for 01005" print applications, Type5 is most preferable. Type 6 is for dispensing applications like jet printing and ultrafine applications.

Alloy powder manufacturers use various types of meshes for segregating particle sizes and classifying the types. Any particle with a nominal diameter of fewer than 1.7 miles will pass through the 325-mesh size (−325) and will catch in the finer mesh size (e.g., +400 or +500). Mesh count determines the size of the opening and hence the particle Size. The higher the mesh count, the finer the particle size.

ASTM Mesh Designation	Opening Size	
	Size (µm)	Size (in)
200	69	0.0027
250	58	0.0023
325	43	0.0017
400	38	0.0015
500	30	0.0012
625	20	0.00078

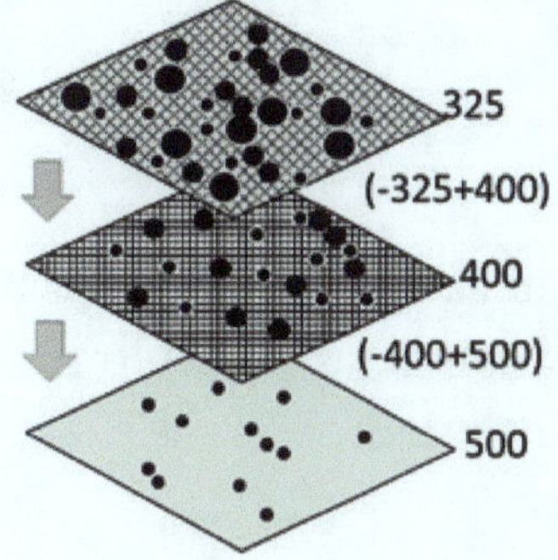

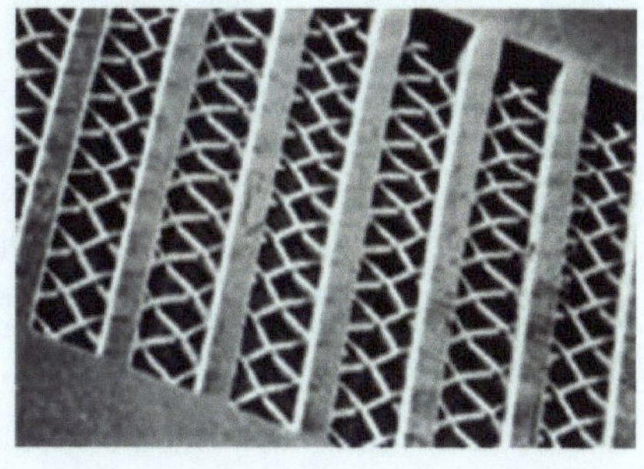

The particle size is generally identified by using two numbers, for example: −200/+325. The correct particle size shall be selected according to the minimum pitch and population density of the fine-pitch components.

The surface area of the solder Powder is also the most important factor because it plays a role in the reactivity of the solder powder. The higher surface area of smaller solder powder types causes the rate of reaction to be higher than the larger solder powder types.

Particle Type	Particle size			Mesh Size	Application
	Mils	Millimeters	Microns		
Type 1	3.0 – 6.0	0.075 – 0.150	75 - 150	-100 / +200	Standard Pitch
Type 2	1.8 – 3.0	0.045 – 0.075	45 - 75	-200 / +325	Standard Pitch
Type 3	1.0 – 1.8	0.025 – 0.045	25 - 45	-325 / +500	Fine Pitch
Type 4	0.8 – 1.52	0.020 – 0.038	20 - 38	-400 / +500	Fine pitch
Type 5	0.62 – 1.0	0.015 – 0.025	15 - 25	-500 / +635	Ultrafine pitch
Type 6	0.2 – 0.6	0.005 – 0.015	5 - 15	-635	Solder Bumping

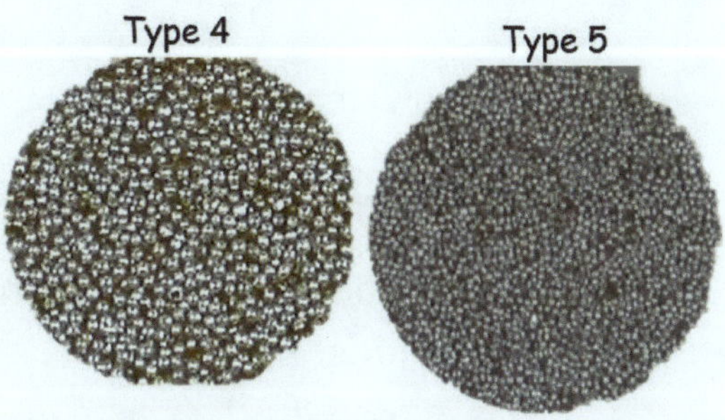

Oxide content should be carefully controlled as low as possible. Smaller solder powder types are more susceptible to oxidation when exposed to air. Particles smaller than 20µm have a high oxidation level. Large numbers of high oxide content particles in solder powder result in the formation of micro solder balls when reflowed. Hence, as the powder size decreases, more flux is required to deal with these oxides.

SOLDER PASTE STORAGE AND HANDLING

Solder Paste is available in different types of containers, and the Applicable type depends on the Application.

- Jars – Paste dispense by hand. The required quantity of paste is dispensed on the stencil from time to time. Paste viscosity measurements more easily if you use a Jar.

- Cartages – Solder paste in cartages dispensed either by hand or pneumatically on the stencil, but widely used for auto-dispensing.

- Syringes – Paste Syringes used in repair environments.

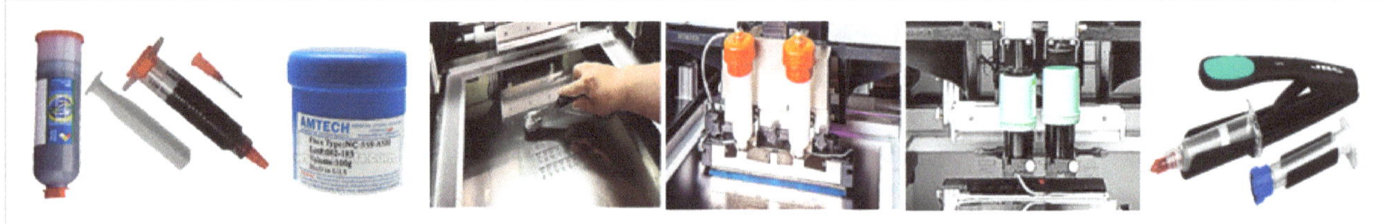

One of the most overlooked areas of solder paste printing is the issue of paste maintenance. Regardless of whether the application is large or ultrafine pitch, proper paste preparation and maintenance are the keys to good printing.

Solder Paste Storage Recommendations

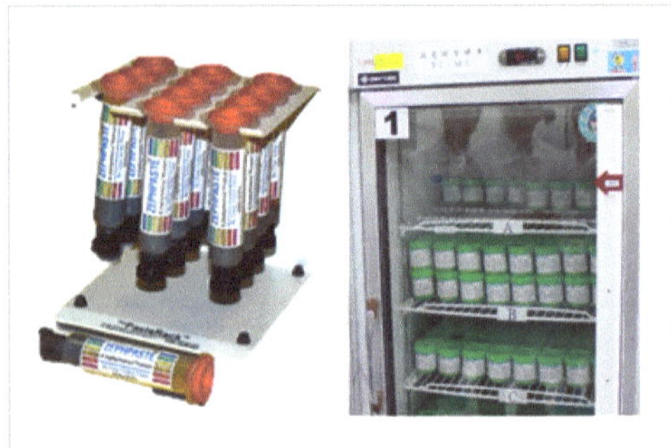

Solder paste shipped cold, –5°C to 10°C

- Refrigerated storage will prolong the shelf life of solder paste, -5°C to +20°C

- Storage temperature should not exceed 22°C to 25°C. Air-conditioned temperatures are usually adequate.

- Syringes and cartridges must be stored tip-down.

- Solder paste is a shelf-life item and should be managed in a FIFO manner.

Process Instructions

- If the solder paste stored at room temp 23.5 °C – 26.5°C, it can be used straight away for printing.

- If stored temperature between 15°C – 22°C, should be used in one hour's time.

- If stored below 10°C, it must be allowed to return to room temperature before opening and use.

- Recommended paste packaged in jars; allow at least 4 hours to return to room temperature before use.

- Mixing of Old and new solder paste is not recommended

- Solder paste performs best when used in a controlled environment. Maintaining an ambient temperature of between 23°C and 30°C at a relative humidity of 30 – 60% will ensure consistent performance and maximum life of the paste.

SUMMARY

The soldering process is performed using alloys with a melting point range of 90°C - 450°C, called soft solder, commonly used in Electronics, plumbing, and sheet metal works. The major alloy element, Tin, is most mixed with silver, copper, gold, Bismuth, etc. Soldering is performed using an alloy with melting above 450°C called a Hard solder. The major alloy element is Copper mixed with either Zinc or Silver, commonly used for Brazing applications.

Tin and lead solders paste alloy ruled the electronic Industry for decades and were almost eliminated from the electronic industry from the year 2000 onwards due to restrictions on the usage of Hazardous substances, especially lead Pb. Still, in some electronic assemblies, especially for Defense and Medical, usage of Leaded Paste is due to its excellent wetting property.

The RoHS (Restriction of Hazardous Substances Directive) rule was imposed effectively in July 2006. Usage of the SAC (Sn, Ag, and Cu) alloy family was introduced for soldering purposes in the electronic Industry. Though lead-free solder pastes showed good wetting and reliable solder joints, higher reflow temperature is still a major concern for design engineers.

Drastic research and development are happening to invent alloys for lowering the reflow temperatures at the same time without affecting the Wetting and reliability characteristics. Developments are in progress to use Bismuth (Bi) as one element where it can replace silver as an alternative.

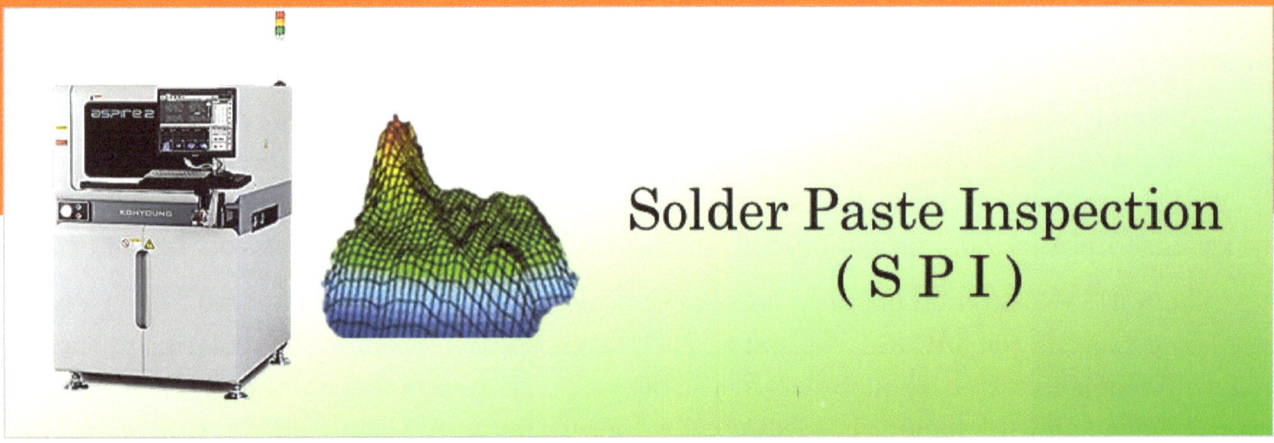

ABSTRACT

Solder paste deposition is vital in board assembly technology in Solder Paste Technology (SMT). This process is prone to defects and errors of approximately 50 -70% during the manufacturing process because of the number of process variations involved (more than 30 variables that need to be controlled). Defects can be either in the solder paste, stencil design, printing process, or Screen-printing equipment. Till the late 1990s, assembly manufacturers used to employ manual Visual inspection on a sample basis to check paste-printing defects with certain predefined critical areas in the paste-printed PCB. That was an era where 60% of through-hole mounting was still popular and used to analyze the defects using the quality concept "vital few trivial many".

Quality engineers used to draw time-consuming Paratoo analyses, then find root causes and take appropriate Corrective and Preventive Actions. Fixing defects was very time-consuming as defect identifications were very subjective, and decisions were made resulting in material wastage and excessive downtime.

From 1995 onwards, demand for electronic applications needed to be more functional while being miniature. For this purpose, the SMT industry and Semiconductor Industries developed various technologies to reduce the number of boards in a product and downsize the boards, driving to thinner and smaller packaging with finer pitch, i.e., small-volume solder paste deposits.

As a result, a reduction in the size of surface mount device packages from 0805 -0603- 0402- 0201 -01005- 03015 to 008004. Similarly, BGA and CSP array pitches reduced at intervals from 1mm to 0.5mm, further to 0.4mm, 0.35mm, 0.30mm, 0.25mm, and still further to 0.20mm. Today's leading-edge mobile phones contain 0.3mm pitch CSPs and chip components 01005s. That means small volume deposits largely below 250 cubic mils (0.025mm =1mil)

INTRODUCTION

When considering high-density mounting, it is necessary to decrease the pad size as much as possible and narrow the gap between the devices. As per the growing demand, technology changes were witnessed in Solder paste, paste composition, printing options, component placement, and reflow profiling.

Shrink in Component size, lead width, and lead pitch, and an increase in package pin counts pushed designers to decrease the pad sizes and need for precise, accurate, and smaller solder paste deposits with tighter tolerances. For PCB assembly manufacturers, profit margins are limited to volume production, and taking the risk of print defects has become a big challenge. Humans intervene on 100% inspection on every Board on the production line, which is not at all possible.

SMD Package		Dimensions, length × width	
Metric	Imperial		
0201	008004	0.2 mm × 0.1 mm	0.0078 in × 0.0039 in
0402	01005	0.4 mm × 0.2 mm	0.0157 in × 0.0079 in
0603	0201	0.6 mm × 0.3 mm	0.024 in × 0.012 in
1005	0402	1.0 mm × 0.5 mm	0.039 in × 0.020 in

BGA➡	µBGA, CSP, PBGA, CBGA, CCGA, CLGA, LBGA, SBGA,T BGA &WLP
Ball Pitch	Fine pitch – 0.5mm, 0.75mm, 0.8mm Standard Pitch – 1mm, 1.27mm, 1.5mm
Ball Día	0.4mm to 0.75mm
Ball count	2x2 (4 balls) To 50 x 50 (2500 balls)
Size	0.6mm To 15mm
QFP➡	TQFP LQFP QFN
Lead Pitch	Fine pitch – 0.3mm, 0.4mm, 0.65mm – 1mm i.e., 11.8 mils TO 40 mils
Lead count	32 leads To 304 leads

Further, with the Increased use of Ball Grid Arrays (BGA), µBGAs, Fine pitch and Ultra fine pitch Quad pack packages (QFPs), Package on package (POPS), QFN, and CSPs, there is a dramatic increase in demand for solder paste inspection after stencil printing. Demand for 100% inspection and online corrective actions becomes mandatory.

So, the question arises: What parameters exactly need to be inspected? Is it the deposited paste area, solder paste volume, paste height, or paste position on the Pad? The answer is ALL parameters. These inspection parameters determine the type of defects, such as Excess solder, less solder, bridging, etc. The accurate measurement of these inspection parameters is not an easy task. Controlling and accurately measuring solder paste volume has become critical as the lead pitch becomes smaller. To identify and remove defects at the earliest possible process step, a 2D and 3D solder paste inspection system should be incorporated to monitor the solder paste deposited on all pads on every board before component placement. There is a critical need for accurate measuring tools such as Offline or inline Solder Paste Inspection (SPI), especially for paste volume deposits.

SOLDER PASTE INSPECTION - PURPOSE

The purpose of SPI is to monitor the solder paste deposits on the pads on the PCB after the solder paste stencil printing process. The purpose of this measurement machine is to detect abnormal patterns developed by the paste printing Machine (Screen printer) either due to the Printer defects or process defects.

Industry research shows that solder joint quality and reliability are directly proportional to solder paste volume deposits on a pad. Too much or too little translates into unreliable joints - an unacceptable outcome. Hence, measurements and inspection by the SPI machine, concerning solder paste alignment and paste volume parameters, should be accurate and need to provide detailed feedback on key parameters of soldering paste deposits, such as Volume percentage (Paste deposit volume Stencil aperture Volume, Area percentage (Paste deposit Area ÷ Area of stencil Aperture opening), paste height, and paste position relative to SMT Pad.

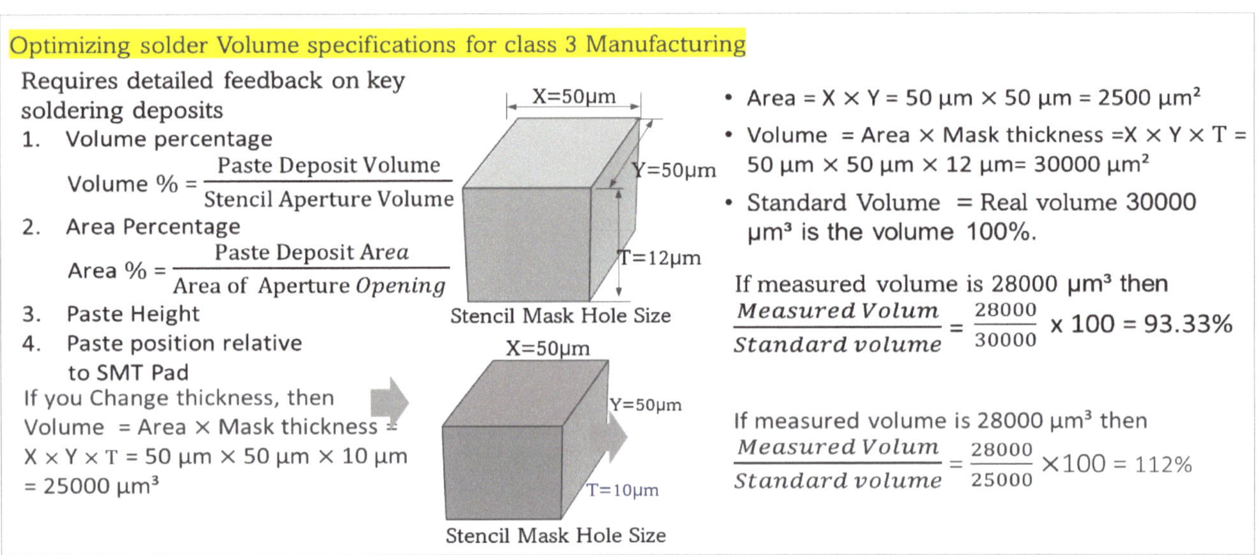

Analyzing the results of Inspection through SPC Statistical process control is another key requirement to beat the digital revolution by using SPC software directly embedded into SPI, making the quality management system operator-free and more analytical in its approach. The SPC software can correlate cross reference and analyze the data in real-time giving immediate feedback on the printing process i.e. Defect location on each PCB, Top 10 defects, real timer monitoring (line performance), 3D defect images, Cp/Cpk and various detailed Pareto and Pie charts which allows or signals process and quality engineers for preventive measures.

SPI – A PASTE MEASUREMENT TOOL

The printing process means more opportunities for defects. Real-time measurement of Solder paste height, paste volume, paste coverage, and alignment of solder paste can predict the quality and long-term reliability of the solder joints. So, an inspection measuring tool (SPI) should consider three factors: 1. Accuracy, 2. Repeatability, and 3. Capability. Other factors such as inspection speed, cost, features, and selection of Inline or Offline SPI can be an option for PCB assembly manufacturers depending on quality commitments.

Accurate measurement of solder paste deposit parameters, either in 2D (two-dimensional) or 3D (Three-dimensional), is a big challenge for SPI OEMs. Challenges such as PCB Warpage, undulation on the PCB surface where the PCB surface is not smooth, PCB color sensitivity, and solder paste deposit not perfectly defined cylinder or cubic shape. Another major challenge for SPI OEMs is paste deposit location, i.e., solder pad / Metal pad or Surface finish. A solder pad is a mix of three intersecting surfaces, i.e., metal pad + solder mask + PCB Material. This complicates identifying and setting up the reference plane for measurement because the exact height of the metal pad for the reference plane is difficult to determine. Small deposits are more sensitive to PCB warpage.

Further measurement variations could be SPI equipment variations, image sensors resolution, operator variations, for example, Height threshold settings, etc.. Measurement process variations, such as the process algorithm that computes height and volume, and even environmental fluctuations.

WHY 3D INSPECTION?

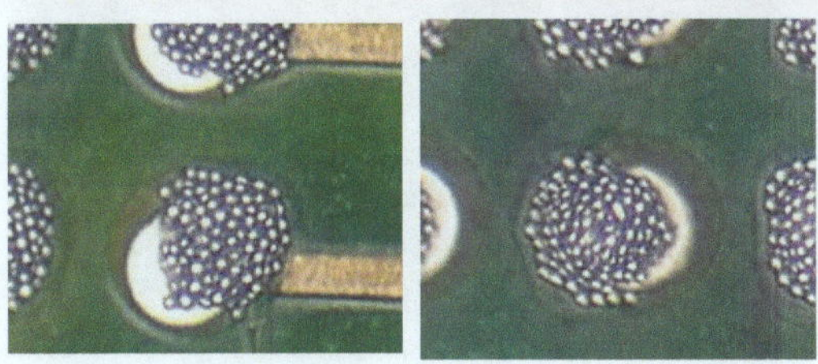

Solder-joint reliability and strength are determined by the volume and shape of solder paste. Parameter Paste height required for Volume calculation. The current market is familiar with fine pitch parts and 0201, Six Sigma Movements, which make 3D (solder paste) inspection necessary for paste height inspection.

1. Area vs. volume

The two deposits look very similar in terms of area. But they are far different values in volumes

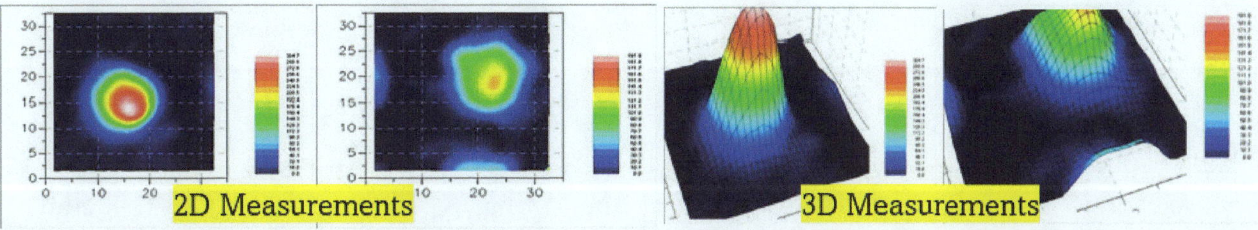

UNDERSTANDING THE WORKING PRINCIPLE OF SPI

SPI utilizes angle cameras that take rapid 3D pictures to measure solder paste volume and alignment. The working principles of 2D and 3D are the same, but 2D measures only the XY offset, whereas 3D measures the XY offset as well as the height. 3D captures the Height of the solder paste on a pad (also known as "Z" for the Z axis), which enables the equipment to measure the total volume of paste printed more accurately.

The core technologies for solder paste Inspection systems are 1) Laser-based and 2) Phase shift or White light Moiré interferometry.

LASER-BASED INSPECTION TECHNOLOGY

Laser-based technology utilizes the core principles of laser triangulation. A laser stripe is projected at a specific angle to the imaging camera. As the laser moves across the surface, a string of data is automatically collected from the Centroid of the laser stripe, generating a differential height reading from the base surface to the overall surface of the solder paste deposit

Triangulation displacement uses a laser beam to measure solder volume without contact. With the triangulation method, each pixel can obtain distance images consisting of height data. This distance image, which calculates the solder volume, is a 3D profile of the printed solder.

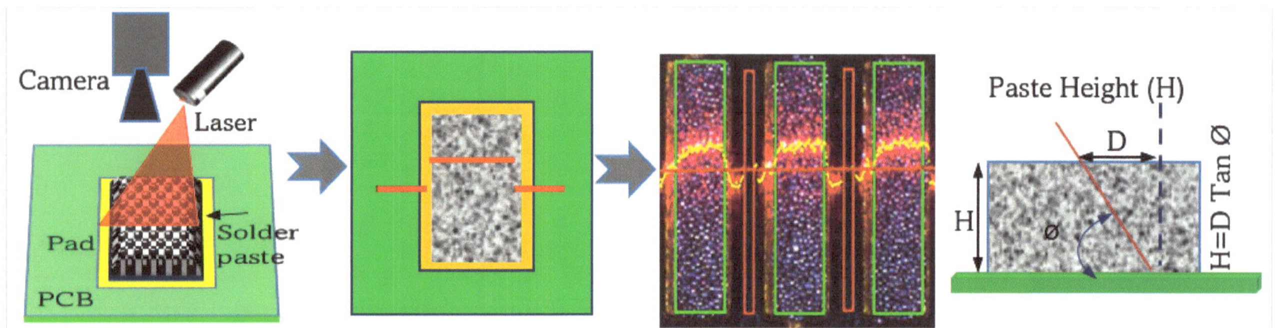

Disadvantages of Laser Technology

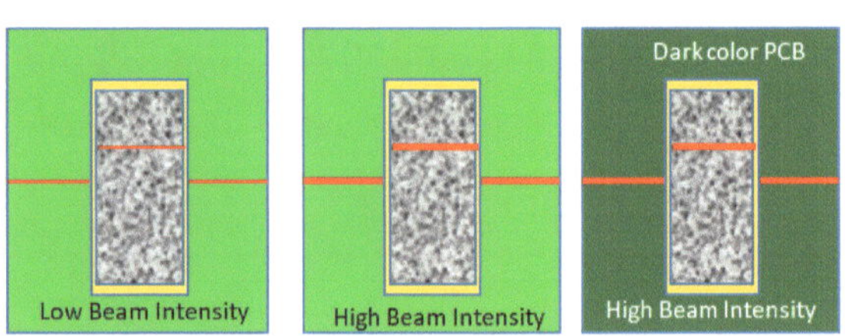

A low-intensity laser beam provides more accurate solder paste inspection results, better focus, and better measurement accuracy but slower. Whereas high beam intensity, brighter line, faster, but less accuracy. Moreover, measurement speed and measurement accuracy vary for dark color PCBs, i.e., they are slower and less accurate. Hence, the Laser Beam Inspection method has a lot of variables concerning beam insanity and PCB optics.

Laser makes a single scan, delivering a single data point per pixel –no possibility to identify if data is good or bad. Laser Intensity can affect the result of the measurement, which may lead to an inaccurate result. The user requires a lot of fine-tuning. Safety precautions are needed due to the laser. Overall performance results - Not accurate, not repeatable, weak against the optical noise, critical for vibration noise, and specula problem.

PHAGE SHIFT OR MOIRÉ TECHNOLOGY

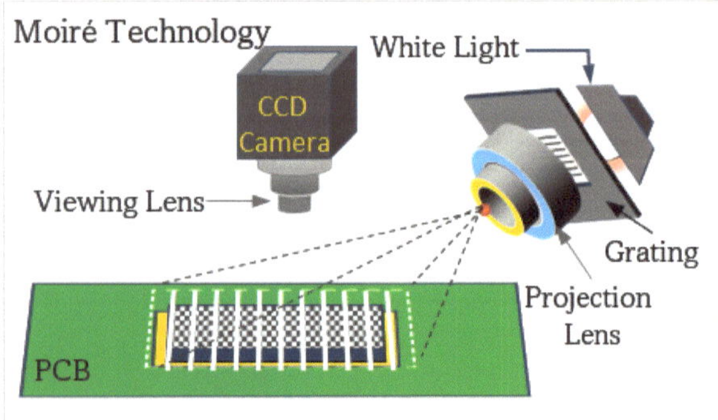

This method uses the concept of stripe projection and white light projection through power LEDs, and dual projection makes the inspection indifferent to changes of colors and materials. This is more accurate and repeatable. No laser safety-related issues.

A white light source projects a pattern of well-defined periodic fringes onto the object (PCB). A two-megapixel CCD camera captures the projected image four times in sequence, including the height information of the object by pixel. As those four frames of captured images are being processed, their height information is significantly amplified by interference between the four image frames and the reference. It enables us to create 3D shape images of all solder paste deposits, measure volume, monitor shape deformity, and so on...

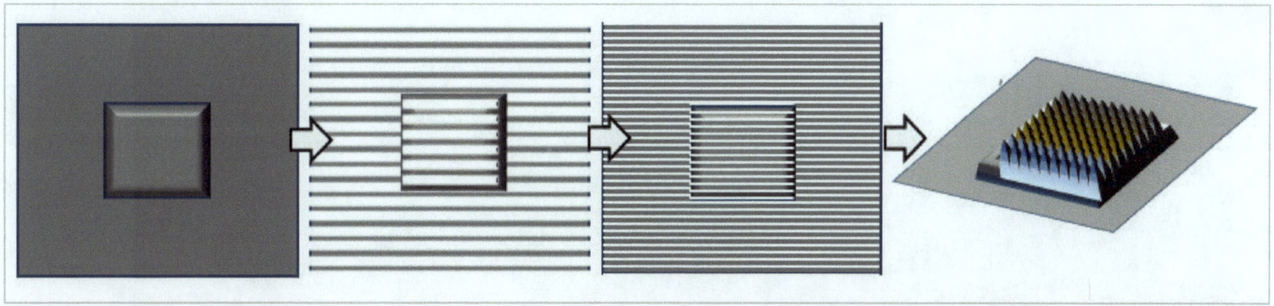

BOTTLENECKS OF THE OPTICAL TRIANGULATION PRINCIPLE

There have been two well-known problems in three-dimensional shape measurement and inspection.

1. **Shadow effect** - Pastes shaped as slopes that are steeper than the projected pattern or light cannot be measured and can only be guessed.

2. **Specular area** - Some pastes may give a specular reflection of the projected pattern or light to the camera or sensor. As a result, we can expect unreliable or noisy 3D measurement data for that area.

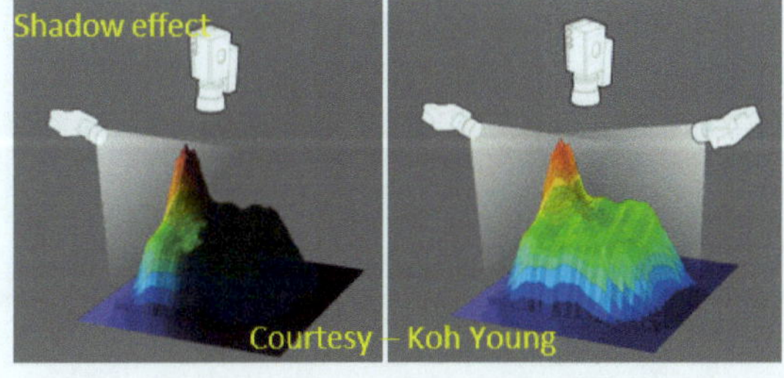

Solution for above Bottleneck - Capture the Right set of the 3D image data consisting of 4 image frames: (R1, R2, R3, R4) and the same way Capture the Left set of the 3D image comprising of 4 image frames: (L1, L2, L3, L4). At each pixel, compute the "S/N Ratio Formula" to get PR and PL;

MEASUREMENT REFERENCE (REFERENCE PLANE)

In general, 3D solder paste inspection machines measure the solder volume, area, and height based on the resist reference or use the pseudo pad reference method.

1. **Resist Reference** - In the case of resist reference, the solder that is lower in height compared to the resist surface will not be calculated in the volume measurement. Since the resist thickness will vary from PCB to PCB, this method cannot measure true solder volume. Trends like higher component placement density and smaller components have led to much thinner stencils at the solder paste print. Resist thickness differences of PCBs now affect the process significantly.

Therefore, solder volume measurement based on pad reference (and not resist reference) is becoming a requirement. Newly developed sensors for paste inspection systems can be referenced from the pad or copper layer. OEMs use their patented technology, which automatically calculates the zero-reference plane using every pad's neighborhood without any programming needed, which eliminates the influence of the board warp.

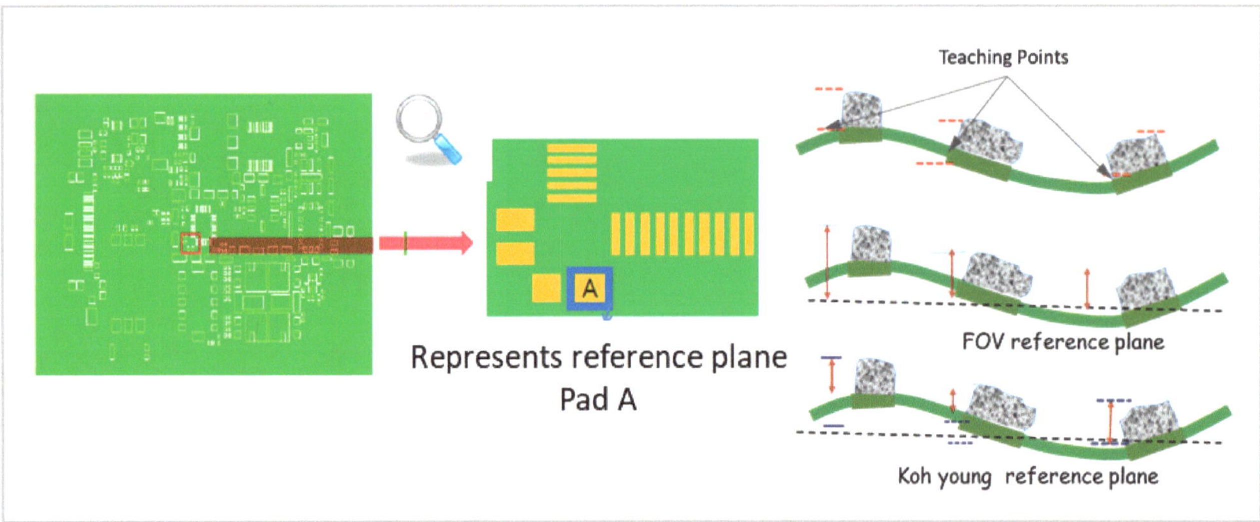

PROGRAMMING

Make programs have an option either Online or offline. Automated solder paste equipment mainly requires Gerber data. The software will take care of Conversion, assignment of fiducials, and inspection parameter set-up. Inspection results & statistics based on PAD numbering. Optional use of CAD Data, combined with the stencil Gerber, which provides data of Inspection results & statistics, has component & pin ID.

SPI – MACHINE CALIBRATION

The greatest aspect of the SPI machine is to maintain its accuracy level and mainly repeatability. Many factors can contribute to measurement variation. Variations can concern the machine itself, Operator-related, measurement process variation, variations among test samples, and could be with Environmental fluctuations.

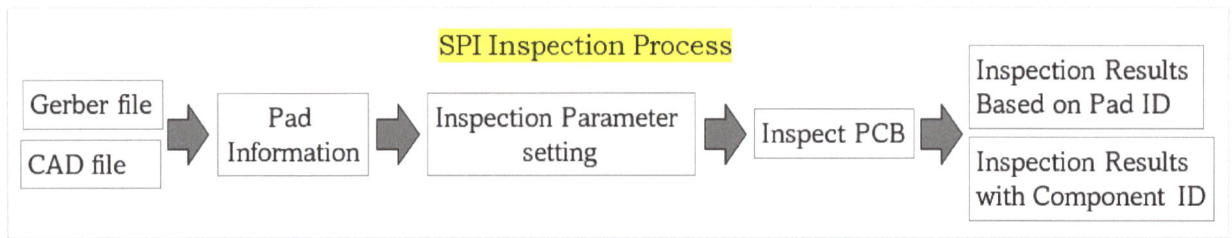

Hence, from time to time, the machine needs to be calibrated using some standard method. Most OEMs develop their own gauge tool and inspection methodology. The Golden reference Board (GB), Non-Solder mask defined pads (NSMD), and Golden reference Tool (GT) are three methodologies in practice by equipment manufacturers. The Golden Board reference tool methodology (GR&R) is the preferred method and is well accepted by OEMs.

GOLDEN BOARD / REFERENCE TOOL

An industry-accepted practice to assess the repeatability of SPI platforms is a Gage R&R study. This golden board is unique from manufacturer to manufacturer but typically made up of a 10-layer, 32 mil thick PCB, consisting of BGAs, micro BGAs, CSPs, micro CSPs, passive, and QFP pad structure. Pad structure for all types of packages mixes Solder Mask Defined pads and Non-Solder Mask Defined Pads. That means, test board featuring 0.76mm(30mil), 0.6mm(25mil), 0.5mm(20mil), 0.4mm(16mil) and 0.3mm(12mil) pitch QFPs and BGAs

SAC305, Type 5 solder paste is used for paste deposit on this test board and dried by heating in an oven at 105 to 125°C for approximately 1 hour to drive off the liquids and solidify the paste deposits to keep the shape of deposit for multiple measurement usage. The board is then run through the SPI platform multiple times, including rotations of 90 and 180 degrees, for measuring paste dimensions many times

GAGE REPEATABILITY AND REPRODUCIBILITY (GR&R)

Gauge accuracy is achieved by calibrating the SPI to a traceable reference standard. That standard can be GR&R Analysis. The objective of GR&R analysis is to identify and quantify the source of measurement variability.

The gauge repeatability is defined as the measurement variations caused by a measurement system measuring the same dimension many times using the same measurement methodology.

The gauge reproducibility is the measurement variation obtained by different operators measuring the same dimension. Gauge reproducibility can be ignored for a fully automated SPI measuring system.

The measurement system, i.e., SPI performance parameters, can be evaluated in two ways. 1. Absolute Measurements and 2. Relative Measurements. An absolute measurement system is calibrated to a traceable reference standard, such as a standard set by the National Institute of Standards and Technology (NIST). In this methodology, the resulting data is verified against absolute Values. Normally, this practice is done by the manufacturers of SPI. All performance parameters can be obtained by this method.

The relative measurement method with GR&R is commonly used in production. Gauge R &R can be evaluated with the assumed average number of performed measurements, but not accuracy.

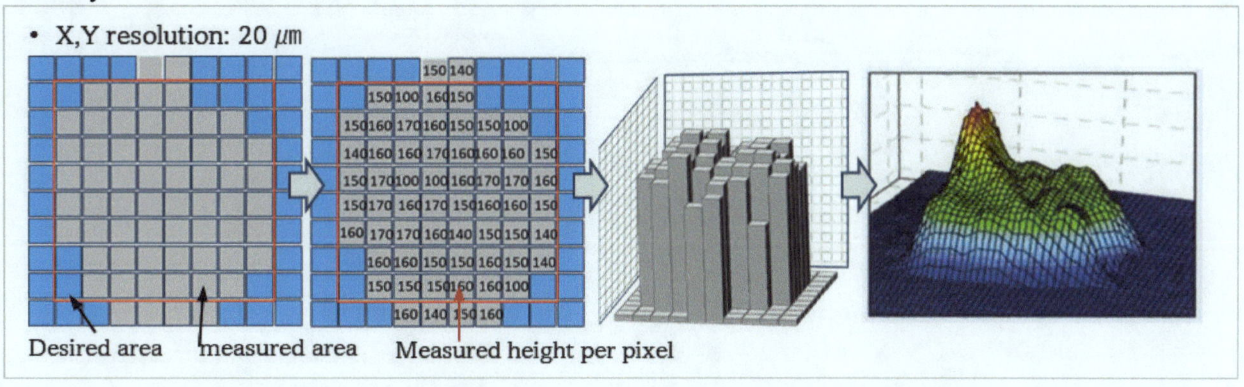

Measured data includes the variance (σ of the product and the variance of the measurement machine in this case, the SPI. Similarly, for Mean μ)(measurement. So $\mu_{Total} = \mu_{product} + \sigma_{measurement}$

- μ_{Total} is the observed average of a set of measurements.

- $\mu_{product}$ is the true mean of the measured product.

- $\mu_{measurement}$ Is the measurement system biased (gauge accuracy).

σ^2_{total} (Overall variance) $= \sigma^2_{product} + \sigma^2_{measurement}$

So a proper measurement system (SPI) should have the capacity to detect variation in the product, which requires that $\mu_{measurement}$ approximately zero, and the ratio of $\sigma^2_{measurement}$ and $\sigma^2_{product}$ should be small.

A simple way of analyzing gauge capability is to use X-bar and R charts. Precision––tolerance (P/T) ratio is commonly used as an Index of gauge capability. The P/T ratio is defined as the ratio of $6\sigma_{measurement}$ to the total tolerance range (the difference between the upper and lower limits of specified tolerance). Generally, a value of P/T of 0.1 or less means it is a proper gauge. Another recommendation is to use the ratio of $\sigma_{measurement}$ to $\sigma_{product}$ OR $\sigma_{measurement}$ to σ_{Total} as a gauge capability Index.

REAL-TIME SUMMARY CHART

All In One page - Example - Glance through this for understanding purpose

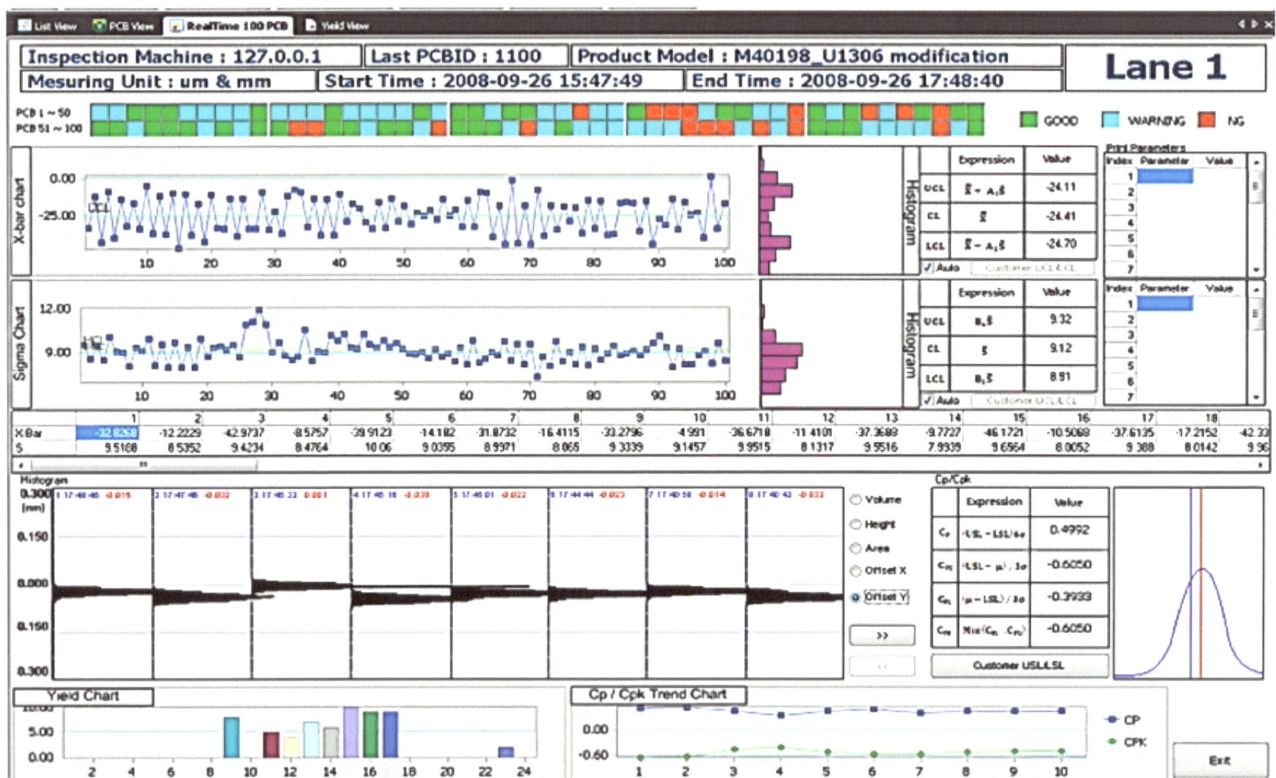

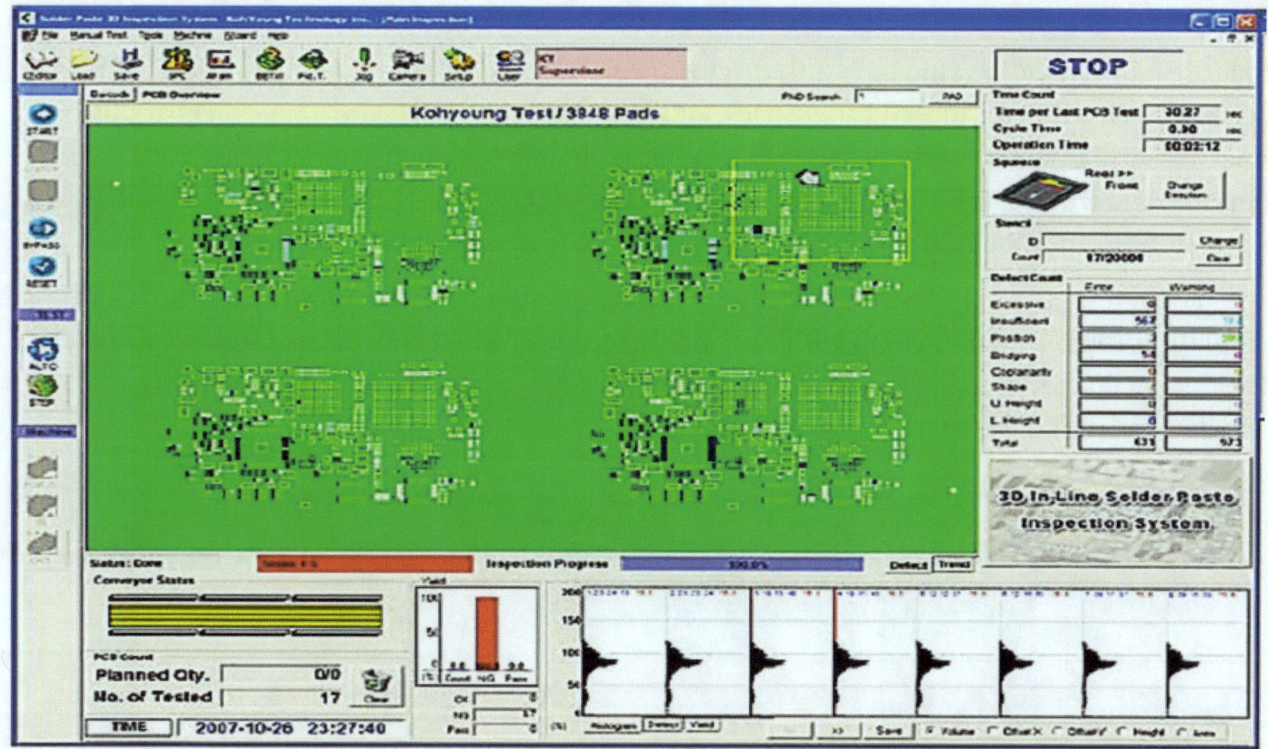

DEFECT DETECTION - EXAMPLES

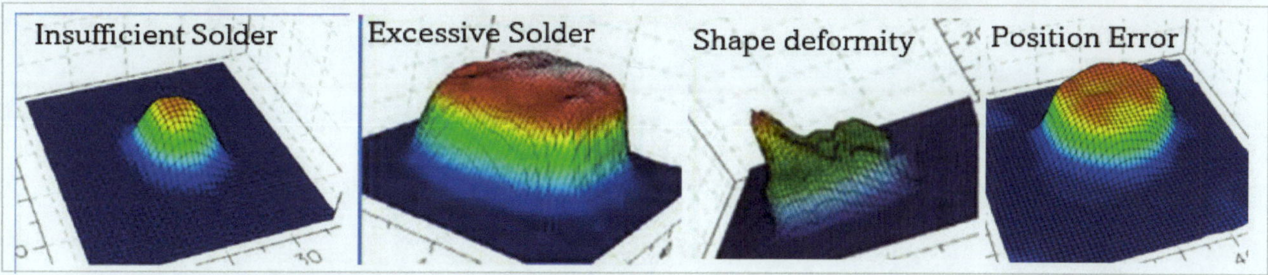

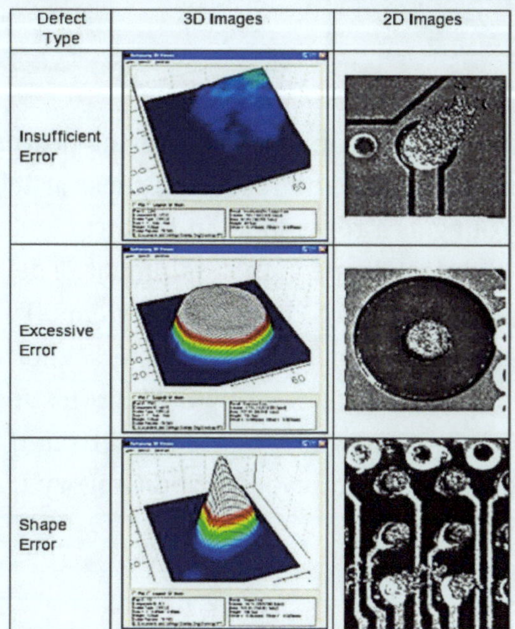

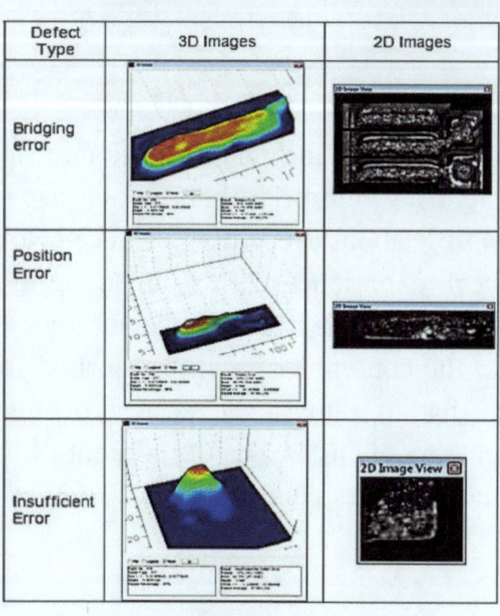

Courtesy – Koh Young

SUMMARY

Smart factory initiation (Industry 4.0) driving the need for development of high-speed and higher accuracy in line SPI for the SMT industries. OEMs updating the SPI technology and introducing in-line Measuring systems with the best SPI measurement performance with << 10% GR&R on 01005 deposits and with <1%@3σ volume repeatability. Introducing True 3D inspection technology with shadow-free, no color PCB sensitivity, and with rapid programming time.

High resolution 4 Megapixel Camera technology adopted for inspection with XY pixel resolution 10µm, 15µm or 20µm with Auto Zoom options. With sophisticated statistical analytical tools, smart factories collect the state data of subsystems using IoT (Internet of Things) sensors and analyze the collected data in real time.

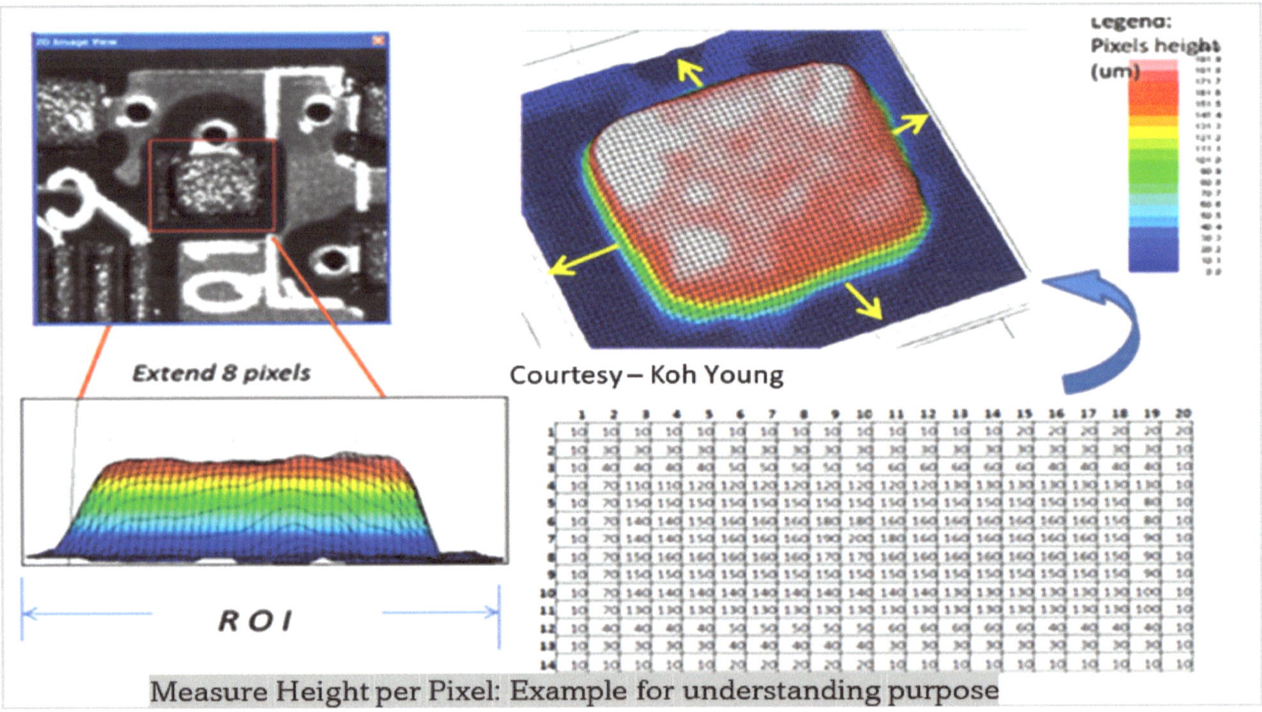

Measure Height per Pixel: Example for understanding purpose

REFERENCE

This technical content includes detailed descriptions of electronic manufacturing processes and the manufacturing enterprise that can help aspirants, students, and readers of these articles come to know "a little about everything". Chosen Koh Young as a role model in this article due to the author's work experience with Koh Young SPI and AOI Inspection Equipment. The relevant content of this report is based on the author's experience with SPI and AOI systems. To prepare this article, the content material is absorbed from various articles, white papers, and technical documents that are published/ released or uploaded on the Internet by SMT professionals and design engineers of OEMs. For the content of GR&R, the material was absorbed from the article "GR &R study for 3D solder paste Inspection by the authors Jianbiao Pan and Gregory L".

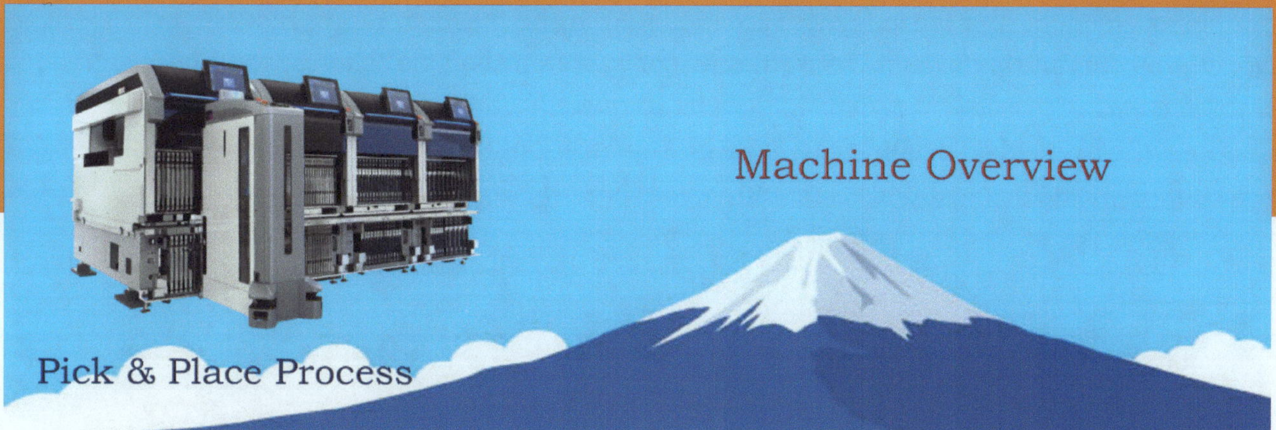

Pick-and-place machines play an important role in the PCB assembly line. Often configured with multiple machines in a production line, the machine quantity configuration will be based on the volume targets, machine capability, and component density. Paste-printed or glue-printed PCBs are loaded into the Pick-and-place machine work area through a conveyor. The pick-and-place machine does exactly what the name implies. Components like Electronic packages, electrical, mechanical, and other relevant components are picked from specified supply feeding locations and mounted on the PCB at specific locations as per assembly programming instructions. The pick and Place Machine helps assemble circuit boards by automating the placement of various surface mount components onto boards. Then, these populated boards are fed into reflow equipment to make intermetallic bonds between the pad and component lead through the solder melt process.

Placement heads in a P&P machine pick the components with the help of vacuum Nozzles or grippers. Picked component it passes on an optical /multi-vision camera that inspects the part geometry and the angle at which the part was picked up, allowing the machine to compensate offsets for final placement. Based on target component characteristics, the assembly program chooses or assigns relevant nozzles to pick from the assigned parts supply location.

Components are fed to the P&P machine either from the front side, the rear side, or both the front and rear sides of the machine. The component feeding capacity depends on the machine's capability. Components are supplied in different packages such as reels, either Tape or Emboss, Tubes, Trays, or, in recent developments, bulk feed. The supply quantity inside the supply package depends on component size.

PICK AND PLACE MACHINE – HISTORY (1980 -2000)

Many types of high-speed pick-and-place machines have been developed since 1980 for PCB assembly. For example, earlier, a typical SMT assembly line employed two different pick and place machines in a line, a Chip shooter and Precision placer, arranged in sequence, to place small and big components, i.e., low and high precision components. Turret-type placement heads are usually called chip shooters, whereas precision placers rely on gantry-type placement heads.

In this model, a fixed turret head rotates/spins, picks components at a fixed pickup location, and performs mounting at a fixed placement location. The parts are supplied through designated component feeders, and those feeders are mounted in the feeder rack at predefined slots. High-capacity motors in the machine move the feeder racks on the axis. The capacity of feeder slots in the feeder rack varies from machine model to model. The feeder rack moves in the axis and positions the respective feeder at the fixed pickup location of the machine, where the placement head nozzle picks the component and moves for the component placement cycle.

The work table carries the PCB through a conveyor and clamps it firmly. In this technology, the work table moves in the X-axis, Y-axis, and Z-axis with the help of respective axis motors. PCB table positions move in respective axes, as per the desired program. Chip Shooter places mainly low-precision components, i.e., simple chip components.

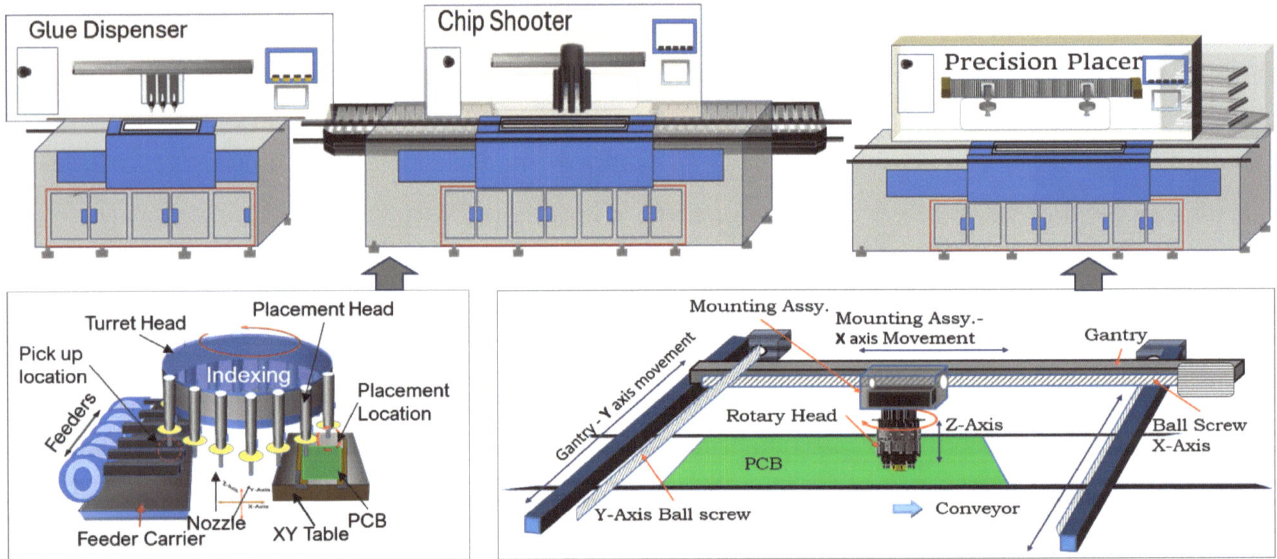

Precision Placers often use high-resolution verification cameras and fine adjustment systems via high-precision linear encoders on each axis to place parts more accurately than high-speed machines. Generally, do not use turret-mounted nozzles and instead rely on a gantry-supported moving head. These precision placers rely upon placement heads with relatively few pickup nozzles. Precision placer places high-precision, critical, odd-form components accurately. Both chip shooter and precision placers of these models were heavy (approx.7000 – 9000kg), and occupied huge floor space (5 -8 meters), which was very difficult to transport and install.

PICK AND PLACE MACHINE -YEAR 2000 ONWARDS

Due to the revolution in component packaging, miniaturization, growing production volumes, and complexity of Boards, pick and place machine manufacturers abandoned fixed turret type chip shooters and precision placers, and came up with various new design pick and place machines such as Compact high-speed multifunction, all in One mount machines, which were occupying less floor space(2 -3 meters), lightweight but stable(2-3 tons) and capable of mounting both chip and precision components.

To overcome the limitations of the electronic component mounting system, gantry-type P&P machines were designed to provide more flexibility to pick and place assembly at higher speeds. The multi-head gantry type machine has a fixed feeder base and PCB work table, which increases the best precision placement. The robot gantry/arm moves the rotary head in the XY direction simultaneously between the feeder base and the PCB conveyor. The rotary head, which is a part of the mounting assembly, moves in the Z direction (up and down) to pick and mount components. This design type of placement head, fixed feeder base, fixed PCB work table, influenced machine dimensions and weight was drastically reduced by approximately 70%.

Fully automated placement systems are categorized into three types: 1. Beam Head 2. Rotary Head 3. Revolver Head. The concept of Beam's head is simultaneous pickup and placement, whereas the rotary and revolver head concepts signify Sequential picking and mounting, which is more flexible. Some OEMs followed the concepts of beam machine layout, some Rotary head design, and some the concept of revolver head type design, placing the head.

The rotary head revolver head machines are designed as high-speed machines. They have similar machine layouts and operation rules, but the only difference is that rotary-head machines rotate nozzles horizontally, while revolver-type machines rotate the nozzle in the vertical direction. Rotary head machines are well accepted by PCB assemblers and are the most popular method in the manufacturing Industry.

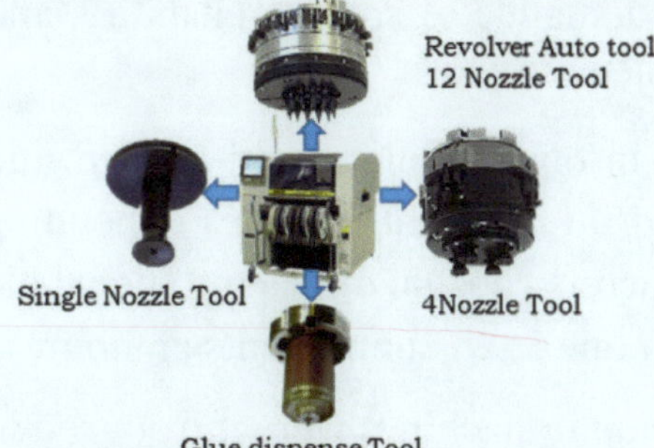

Fixed Head with exchangeable Tool Station
Source : Fuji

Components are fed through Feeders in the Pick and Place machine and feeders are assigned to feeder slots on the feeder base. Selection of feeder type based on the component size it is going to handle. Similarly, Nozzles are assigned to the Nozzle holder and the selection of a type of nozzle-based component handling capability. The feeder arrangement sequence on the feeder base and the sequence of Component pick-up with nozzles are purely based on programmer skill and machine auto-optimization.

The Gantry, also called an XY robot, moves the rotary head towards the feeder base, picks the components through nozzles as per the programmed sequence, and places components on the PCB on the Conveyor after the inspection process. Once the placement is completed, the gantry moves the rotary head back again to the feeder base for the next pick-and-place cycle. This process continues until all the components are placed on a board.

NEW GENERATION PICK AND PLACE MACHINES

Machine manufacturers came up with a new generation of compact, high-speed, versatile placing machines. High-speed multipurpose mounter with very good placing accuracy, capable of packaging Tape or reel parts, tube or stick parts, tray parts, and feeder bases on both front and rear sides. All in one concept. No need to follow conventional line configuration such as glue machine, chip shooter, and precision placer for the assembly process.

A single machine can perform the placement of low-precision components, high-precision components, and glue dispensing with a concept of dynamic head exchange. Head exchanges the tools as per the program, selects the Revolver auto tool (12 nozzle tool acts as chip shooter), Single nozzle tool for placement of high-precision components (acts as precision placer), and Glue auto tool (acts as a glue dispenser).

Phase by phase, drastic changes happened in the electronics industry throughout the world. Change in Circuit design, component miniaturization, and design developments in semiconductors, design for manufacturability. Growth of electronics is observed in all sectors – medical, aviation, entertainment, space, and defense, and especially with the intrusion of smartphone technology, there is a tremendous increase in electronic manufacturing company establishments to meet the demand.

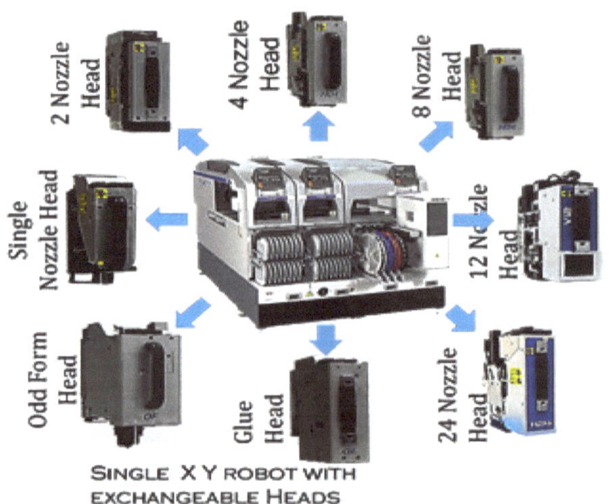

SINGLE X Y ROBOT WITH EXCHANGEABLE HEADS
SOURCE : FUJI MACHINE MFG.CO.LTD.

Hence, equipment manufacturers and their ancillary co-product developers are bound to come up with new technology to pick and place machines' demands of speed, reliability, and capabilities, which made a revolution in surface mount technology.

Machine manufacturers produced new techniques - an all-in-one modular, multi-headed, and multi-gantry machine that could have heads quickly swapped on different modules depending on the product being built, to machines with multiple mini turrets capable of placing the whole spectrum of components with theoretical speeds of more than one Lakh components an hour.

Scalable placement platform (Modular technology) placement machines were introduced in the market, where placing heads can be exchangeable.

Dual Gantry Machines have gained popularity at present. In this technology, the first gantry places the components on the PCB, and the rear gantry picks the components up in the feeder slot concurrently. Dual Gantry machines have the asymmetric layout of two feeder slots, two cameras, two automatic nozzle changers, and two gantries with exchangeable head facility. Even some machine models have 3 or 4 gantries as a special application. Operational modes are classified as Independent, Single dependent, and Dual dependent modes. In Independent mode, two gantries assembled two PCBs independently.

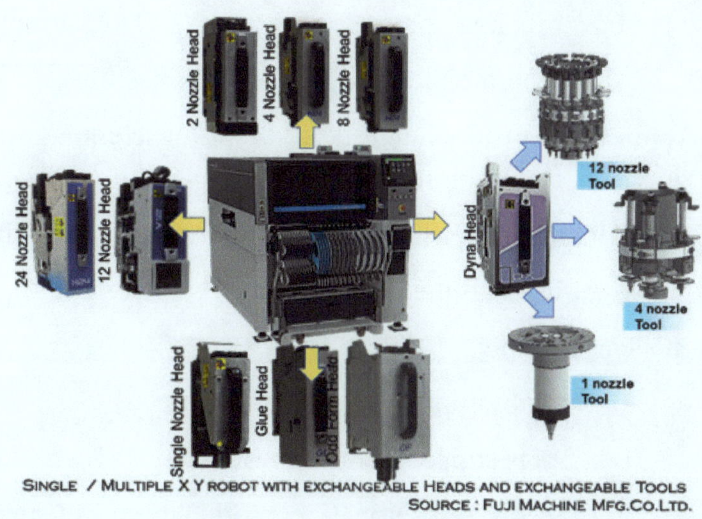

SINGLE / MULTIPLE X Y ROBOT WITH EXCHANGEABLE HEADS AND EXCHANGEABLE TOOLS
SOURCE : FUJI MACHINE MFG.CO.LTD.

Machines are made up of high-accuracy, high-strength hard parts and assembled with high-precision, durable parts that can maintain stable placement quality. Electronic Intelligent Feeders were introduced in place of mechanical and motorized feeders. High-resolution parts cameras, side cameras, and newly developed fly-on Vision parts cameras were introduced.

With the advent of digital engineering technology, mechanical/ control integration technology, linear motor development technology, and developments in thermal and fluid design, placement machines used to develop high-degree accuracy, the hardware used powerful compact linear motors. With more focus on software applications for the process, multi-job line balance focused on the machine's versatility to deal with short runs and fast changeover. The industry is moving beyond conventional component placement with new applications like POP and wafer placement on the substrate.

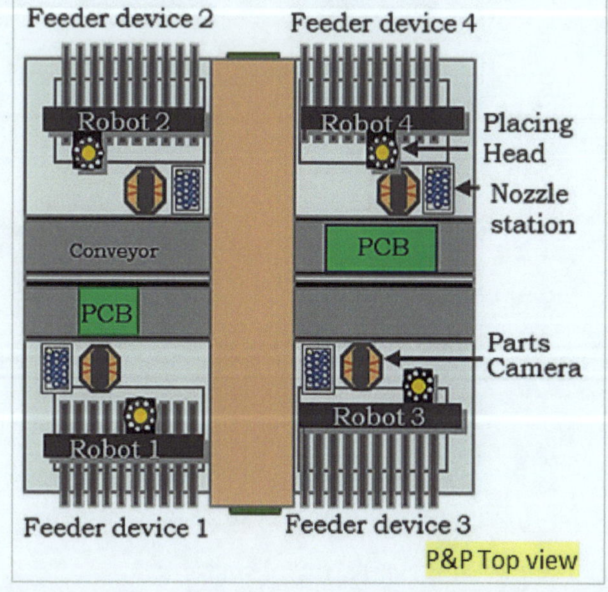

PICK AND PLACE MACHINE MODELS

SMT Placement equipment is available for every level of production, and the selection of equipment depends on Capability, features, durability, adaptability, and finally pricing.

- Proto Typing production

- High mix low volume production (Frequent job changeovers)

- Low mix low volume production (LED Assembly)
- High volume production (Mobile Assembly)

To meet the production requirements, efficient equipment can be selected based on machine configuration, throughput, and parts supply. Equipment selection can be configured based on

1. Machine dimensions (To meet Floor space requirement)
2. Meet Board dimensions (Panel)
3. Parts supply
   - Parts supply from the front side
   - Parts supply from the rear side
   - Feeder Capacity (Capability Number of supply feed)
   - Type Component packages the machine can handle
- Placement Head
   - Requirement of the number of robots to place components
   - Placement speed CPH (Comp/hour)
   - Placement Odd form component
   - Glue Dispense Application
- Production Volume
- Production Changeovers

PICK AND PLACE MACHINE OVERVIEW:

Typical overview of pick and place machines having the following components

- Parts supply system
- Parts mounting system
- Vision system
- Panel conveyor system
- Electrical control system
- Pneumatic control systems
- Other optional components

Machine Aesthetics and technology can be different from machine manufacturer to manufacturer, but the machine layout will be the same for all. Menus can be different, but basic operations are always the same.

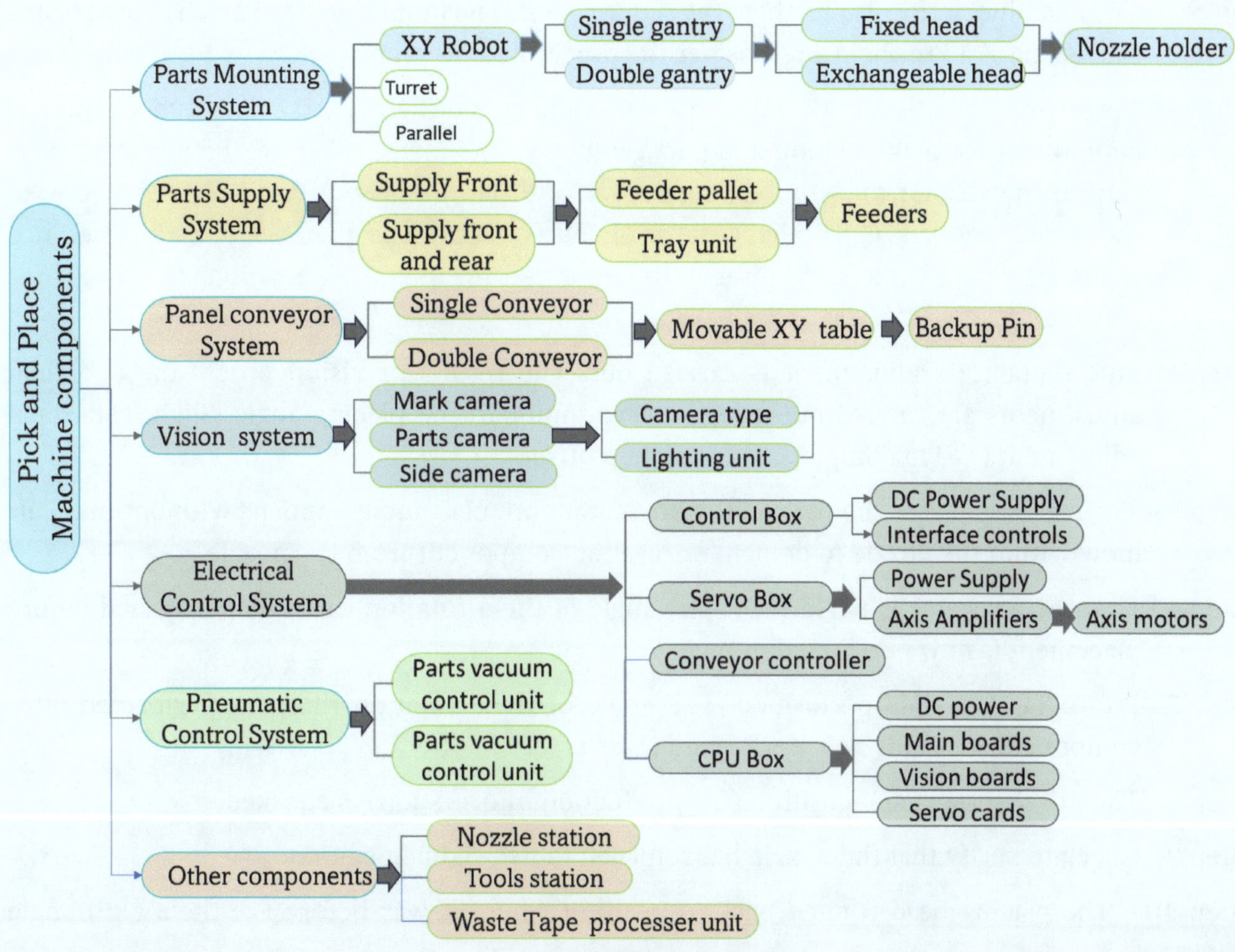

WORKING PRINCIPLE - PLACING HEAD

The work method of placing head is quite simple to understand. Motions are controlled by axis servo motors and all servo motors are controlled by encoders and servo amplifiers. The Gantry X Y robot moves in X and Y directions with the help of X-Y axis motors, and the placement head moves along with the XY robot to complete the cycle of Pick and place tasks. The rotary head prompts the nozzles to move up and down, i.e., the Z direction, for the component pick and while placing. This action is controlled by the Z axis servo motor. To pick the components by all associated nozzles, the rotary must spin in sequence in one direction to pick and place the components. This action is controlled by the Q axis servo motor. As per the placement program, picked components need to be rotated and placed at a desired location on the PCB. This rotation action is carried out by the R axis servo motor. The placing head performs several special functions repeatedly, in sequence, for every pick and place process cycle.

Step 1 – Nozzles in the head assembly pick parts from the feeder as per pickup sequence. Pick up feeder location and quantity per location as per user-determined program. Immediately after component pickup, the feeder mechanism advances the feeder tape so that the next part is ready for the next pickup.

Step 2 - The machine performs pre-theta ($\rho\theta$) adjustment. The mounting head mechanism rotates the nozzle at a Pre-theta angle as per the predefined program.

Step 3 - Vision processing is carried out with a Mark camera/side light. Parts Camera checks picked component for pick position and part geometry. The image taken by the part camera is sent to the vision processing for inspection and the processed results are verified with the stored part data. CPU calculates the offsets (offset values lie in between the tolerance limit's part data) values and directs the servo unit for the target positions of the axis for placement.

Step 4 - Fine theta ($F\theta$) adjustment is carried out. The results of vision processing and final adjustments are carried out using a servo-motor to the placing angle. Slight rotational adjustment ($F\theta$) to compensate for pickup offset.

Step 5 - The machine places the part on the board. Smooth placement happens with optimum air blow within the nozzle to detach/release the component from the nozzle.

Step 6 - Fine-theta reverse is performed. The angle of theta rotation that was performed before placement is now performed in reverse.

Step 7 - Pre-theta reverse is performed. The angle of theta rotation that was performed after component pickup is now performed in reverse.

Step 8 - The nozzle holder's A-position (for production information) is checked.

Step 9 - Check to verify that the nozzle has returned to its original position.

Step 10 - The placing head rotated back to position where it was before Pre-theta ($P\theta$) angle rotation.

Step 11 - The part that failed vision processing (and was therefore not placed) is discarded in the dump box.

Step 12 - Nozzle change performed – auto nozzle changes from nozzle station as per placement program.

PICK AND PLACE MACHINE – A LITTLE ABOUT ELECTRICAL CONTROL SYSTEM

A typical electrical control system is quite common for all placement machines and it consists of major electronic circuitry -module CPU box, Vision processing, multi-axis servo packs or boards, Input /Output circuitry boards, and sensors/ relays.

1. Control Box

CPU, The Central Processing Unit can be thought of as the "brain" of the machine, whereas the control box can be thought of as having the "nervous system" of the machine. Signals from various parts of the machine come into the I/O boards, and these signals are then processed through the CPU.

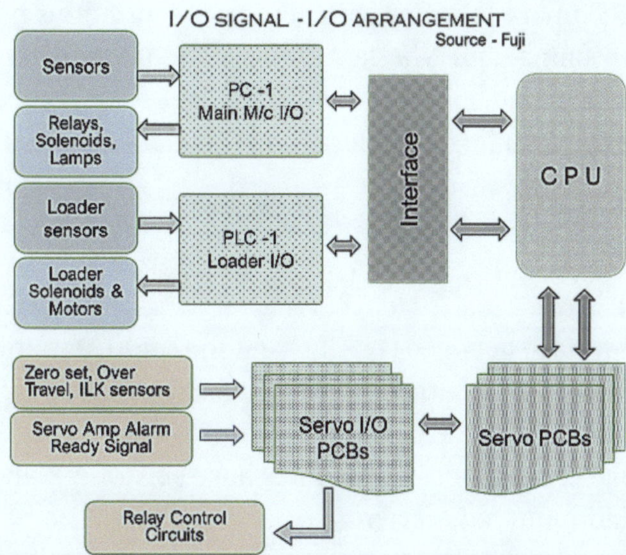

2. Module CPU Box

CPU functions as the brain of the machine. The firmware CPU box contains a set of Servo boards, Vision processing Boards, DC Power Supply units, and CPU Main Boards. The module CPU box takes complete control of Vision processing, i.e., in and out of Part Camera, mark and head camera, display of operational panel, servo actions like all axis amplifiers, and communications. Similarly, the main CPU board controls the functions of the feeder pallets, conveyor controls, and placing head unit. The console board in the CPU Box receives the signals from the CPU and converts them for display on monitors.

This CPU main board carries the processing required for operating the machine as well as storing the programs. It contains RAM where the Proper Data, Programs, and Part data are stored. During the machine's power-off conditions, this data is retained by using the battery. Handles ports such as printer port, RS-232C port, and trace data port.

A set of interface cards is also a part of the module CPU box, which communicates the Input / Output signals between PLCs and CPU, and between the multi-feeder unit (MFU) and CPU, which receives the " delivery complete" signals from the feeder Units. Also, the CPU communicates with Offline Programming devices through data cable RS232 and some other options such as Handy terminals, Barcode readers, and other optional peripherals.

MP Board is one of the PCB cards in the module CPU box where machine firmware is saved on this card by the Manufacturer. The CPU utilizes the board as its memory board. The CPU recalls this data during the machine initialization sequence. This Board also monitors control power AC 100V (emergency stop button) to stop the machine. CPU uses the battery power source to retain the Proper data, status data, and program data when the machine is turned off.

VISION PROCESSING UNIT

A set of vision processing boards controls the vision processing of components that are inspected before placement. The fiducial mark camera, parts cameras, and side camera(optional) are connected to this board. The image taken by the cameras is sent to this board and is verified with the part data. Offsets, positioning, and whether the part should be placed or not are sent to the CPU board. Every time the machine boots up, the Vision Processing board receives the necessary data about vision software for both image cameras and the Mark camera from the vision memory board.

MULTI-AXIS SERVO PACK /BOARDS

Servo boards serve as an interface between the CPU and the servo amplifier. This board interfaces multiple axes in the machine, such as the X axis, Y axis, R axis, Q axis, and Z axis, etc., and several axis requirements depend on the machine firmware design. For example, the latest model modular, head exchangeable pick and Place machines have separate X and Y-axis servo packs for each robot (X Y Robot) and multi-axis servo packs for each robot which controls all other exchangeable head axes R, Q, Z. etc. and the axis count depends on placement head model and internal design.

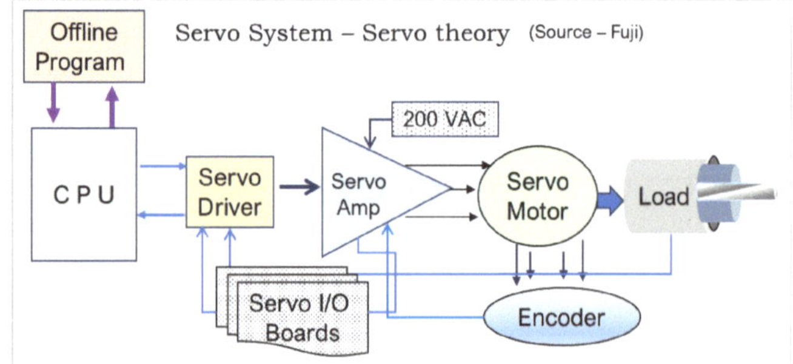

The servo axis board receives data regarding the target position of the axis from the CPU. Then, it calculates the speed and the rotation direction of the motor and sends this data to the servo amplifier.

The Servo amplifier controls the heart of the control system. Motion control terminology of drives. With the help of digital servo amplifiers, more features such as position control, Network communications, digital Input/output, and preprogrammed moves became possible.

All servo motors are AC servo Motors, and motor design defines AC Motor + Electronics to drive the motor + a sensor to measure the position of the motor + a controller which calculates how much power to apply to the motor to get it to the correct position and fast. The encoder or Pulse generator provides the signals 1. Location (i.e., distance in Pulses), 2. Frequency (i.e., speed) and 3. Phase rotation (i.e., direction). The Servo Box has all the servo amplifiers for the servo-controlled axes. These amplifiers directly control the actions of the servo motors.

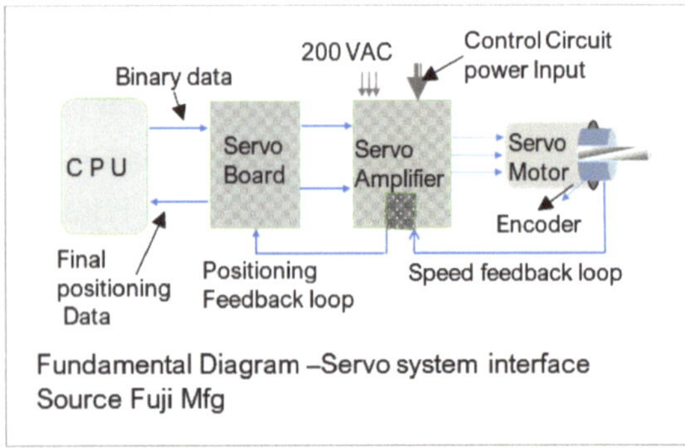

Fundamental Diagram –Servo system interface
Source Fuji Mfg

SUMMARY

Programmers can operate the production lines without stopping the machine during production. Newly developed machine operating systems perform real-time production analysis, multi-job line balancing where production involves multi-jobs, feeder allocation can be optimized for the jobs as a group, and traceability to improve productivity. Improvements made for external changeover (multi-feeder Units, Tray units, reel setting stands) for faster production changeovers. Phase-by-phase machines develop with compact designs with a reduced length.

Present generation pick and place machines can handle very small 0201mm (0.008004"), i.e., 0.25mm X 0.125mm parts with placement accuracy of ±0.025mm. Places components at high speed using low-impact placement and non-tress mounting. The introduction of Faster X-Y robots, faster tape feeders, faster Servo controls, faster vision processing analysis, and sophisticated machine software made pick-and-place technology change towards a smart factory platform. Complete automation is the main agenda for the Smart Factory platform. With the introduction of AI (artificial intelligence, tools were developed that support machines from material preparation to performing maintenance. SMT production from planning and production to management and analysis to optimize the entire factory.

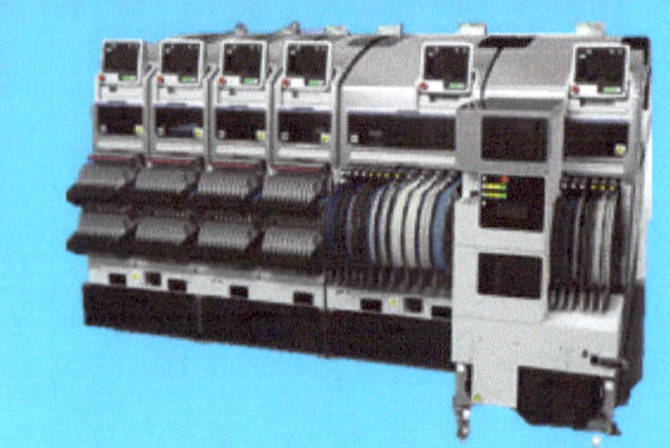

Parts supply system in the Pick and place equipment is one factor that defines its capability and flexibility. Ideally, one can expect that these types of equipment can pick and place all varieties of electronic components such as active components, passive, all types of ICs, and even all families of electrical, and mechanical components. Parts package sizes vary from small, medium, large, and odd forms. The components that need to be assembled on a particular product, i.e. PCB, are fed to the P&P machine so that the machine can pick the components from its pick point and the parts set position and place the component on the desired coordinates on the PCB as per programmed instructions.

Component suppliers supply semiconductor and electronic parts in tape form, trays, bulk, and Tubes. For larger ICs, some electrical and mechanical components are supplied in trays, where the parts are stacked in tray compartments. small components, such as surface mount devices, are provided in chipped tape reels or embossed plastic reels.

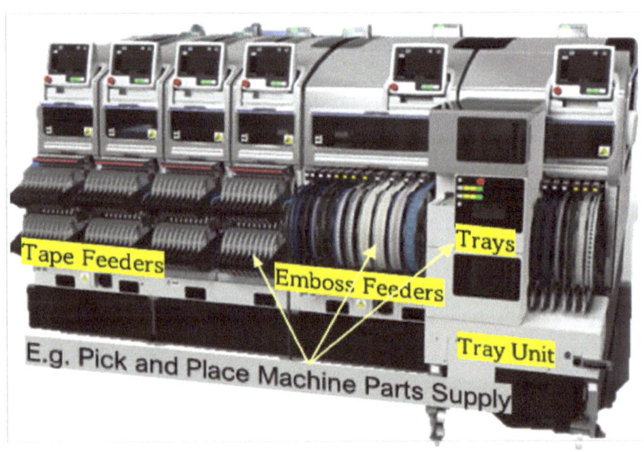

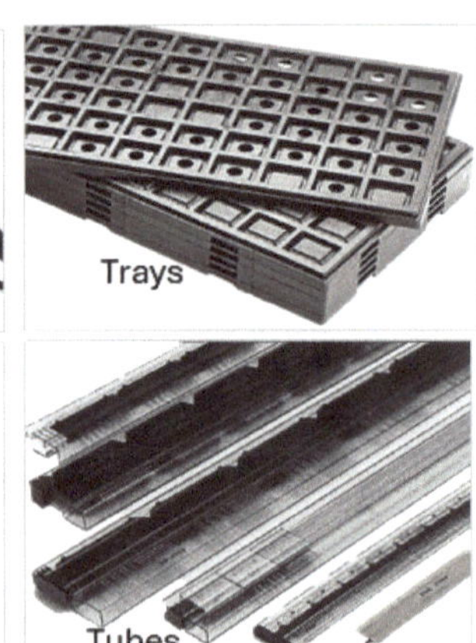

Manufacturers of components, or suppliers and distributors, follow standards like EIA (Electronic Industries Association), JEDEC (Joint Electron Device Engineering Council). These standards provide standardization between manufacturers and purchasers, eliminate misunderstandings between product designs. Package manufacturers and suppliers follow their recommended specifications, such as, let us consider reel as an example, reel size, the width of the carrier tape, and the spacing between components, etc. Pick and Place equipment manufacturers design their component feeding section of the machine based on these standards.

TAPE–ON – REEL PACKAGE

Components supplied through a reel is the most preferable method and economical. Component handling, storage, and transportation are easy when compared with trays and tubes. Easy to store and feed in the pick-and-place machine. Components are supplied on a tape-on-reel in a 15-inch, 13-inch reel diameter, and a 7-inch reel diameter. Small components, either in paper tape or emboss, which can fit in 8mm tape width or 12mm tape width, are supplied in 7-inch reels. Reel Diameter 13-inch and 15-inch can handle tape widths 8mm, 12mm,16mm,24mm,32mm,44mm,56mm, 72mm, and 88mm. Component capacity per reel depends on the size of the component.

One key point to remember about emboss reel carrier tape material and even tray material is that, in some cases, as a part of the process instructions in the PCB assembly line, Reels or trays need to be baked, especially for (HSC) humidity-sensitive components. A typical dry temperature for the Emboss reels is around 40^0C (104^0F) and for the trays 150^0C. Hence, the standard type material cannot withstand high temperatures. JEDEC standards provide guidelines on baking requirements for moisture-sensitive parts.

TRAY PACKAGE

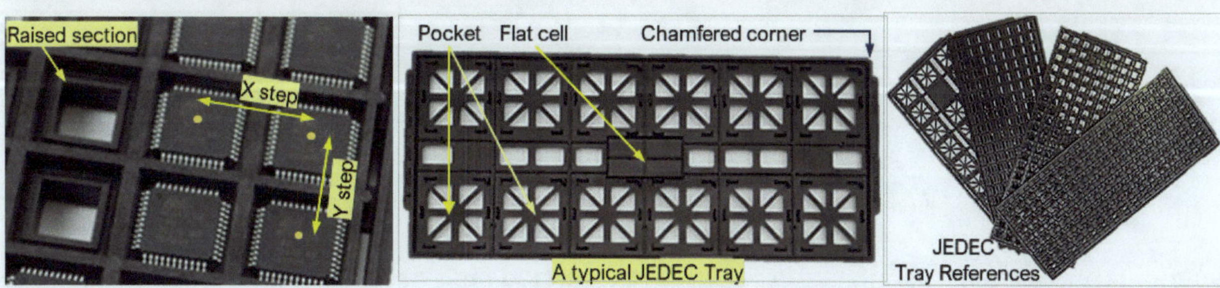

The present trend of IC shipping in Trays became limited to odd form Components, Gull wing lead components such as QFP types, and humidity-sensitive components, as trays can withstand high temperature baking. For other category ICs, such as Ball grid components, etc., end users prefer to order components in reel form as it is easy to handle and more economical.

Integrated circuit devices are shipped in specially designed trays that align to JEDEC standards. JEDEC standardized or assigned tray references for various devices to be stored based on the type of device, body size, thickness, and other geometries. Accordingly, the number of columns, rows, and step sizes (from pocket to pocket) in the horizontal and vertical directions are tailored to the tray material options.

Matrix IC trays are meticulously crafted to meet the demands of modern electronic packages. The outline dimensions of all JEDEC matrix trays are standardized to 322.6mm X 136mm. The maximum number of devices or parts a tray can hold depends on part geometry, tray manufacturing limitations, and application process requirements. The chamfered corner (45⁰ angle) indicates orientation, i.e., a visual indication of pin no. 1 of the IC.

There is a flat cell in the center area of the tray to allow the pick-and-place machine to automatically lift and remove the empty tray from the top of the stack with the suction cup of the robotic arm. If the tray is for a component with pins, there is a slightly raised section in the middle of the pocket so that the body component rests on the raised section and the gullwing leads of that component float in free space and do not press against the tray.

FEEDERS

A feeder is a device that locks tape and reels SMD components, peels off the tape(film) cover on top of the component at each indexing, and feeds uncovered components to the same fixed pickup position for the pick and place machine pick up. Old generation feeders are mechanical, motorized, and pneumatic, no longer in the industry, and cannot fit with modern technology pick and place machines. Improvements in feeder technology mean that Tape format is becoming the preferred method of presenting parts on SMT machines. Present technology limits the supply of Tray parts to odd-shaped components. As the technology rapidly increased, innovations in feeder technology drastically changed in terms of indexing, pick-up accuracy, speed, and repeatability for placing components smaller than 01005mm.

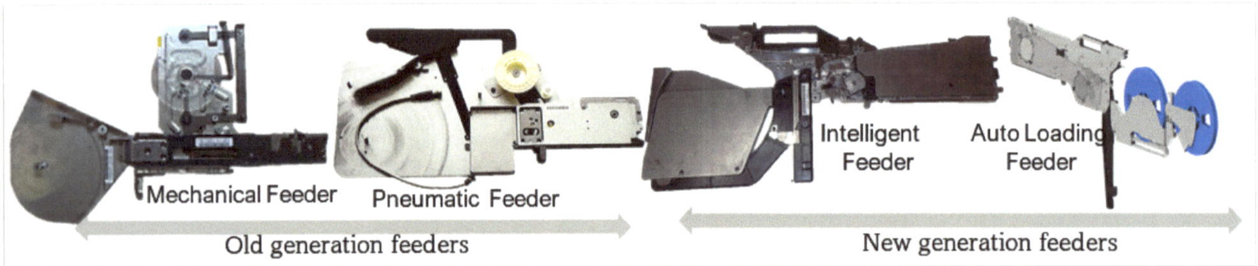

Original equipment Manufacturers (OEMs) developed high-precision calibrated feeders with new index systems, superior gears, encoders, and a control board that can store all the information to meet the demands of traceability. It is possible to set combinations of many types of supply units. It is also possible to exchange units for changeover.

Intelligent Feeders

As the name implies, unlike the old generation mechanical feeders, intelligent feeders contain a control PCB, precision sensors, superior gears, and sprockets. Each feeder has a Unique Identity.

It comprises a communication device at the end of the feeder, which communicates with the pick and place machine, like handshaking, sharing self-information, etc., upon being attached to the P&P machine.

Intelligent feeders improve productivity by enabling a nonstop supply of parts and preventing incorrect part setting. A single feeder supports different Tape packages, i.e., paper, embossed, and thickness, with adjustable index pitches. These feeders can be set or removed during production without stopping the machine. Tape splicing (attaching another reel without stopping the machine to reload) can be done easily, and a splice detection sensor. Motor driven with smooth Indexing, with high stopping accuracy. The feeder positioning section is very close to the pick-up position. Having a mark for offsetting the pickup position. Depending on the component package, corresponding feeders are available in the range 8 mm, 12mm, 16mm, 24mm, 32mm, 44mm, 56mm, 72mm, and 86mm.

Stick Feeders or Tube Feeders

Demand for parts supply in tubes declined as part shipping in carrier tape and Trays is more convenient and economical. End users usually order tube parts for research and development purposes or for the pilot production of new products where production volumes are very low. Stick feeders developed to feed tube parts in pick-and-place machines

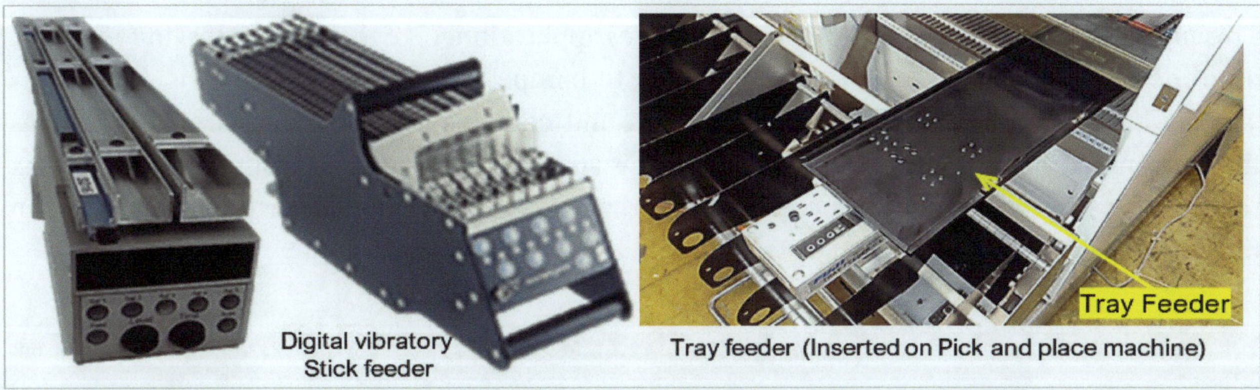

Digital vibratory
Stick feeder

Tray feeder (Inserted on Pick and place machine)

Stick feeders come in different options, including Stick holders with part packets in sizes small, medium, large, and extra-large. Digital vibratory stick feeders with an auto-tuning facility result in smooth and precise feeding of components. For some applications, other types of feeders, such as Horizontal stick feeders, Slop stick feeders, Sky Slope feeders, and Stacked Stick feeders, that work on some form of electromagnetic vibration to move components along to the pick position, are also preferable. The intensity of vibration can be controlled. In the same way, Tray feeders preferred low-volume productions.

MULTI TRAY UNIT: MTU

A multi-tray unit is another option, supplying parts (parts shipped in Trays) to the pick-and-place machine. MTUs play a major role in high-volume PCB assembly production lines. MTU product design is unique, specific to Pick and Place OEMs. Multiple models of tray units are developed by manufacturers based on the size specific to the machine. Depending on volumes and specific requirements, end users can choose specific models for their pick-and-place machines. Multi Tray Unit can be mounted alongside feeders at the machine.

Compatibility varies from Model to Model. Choosing models based on several independent magazines, the number of drawers in each magazine, and the tray capacity in each drawer. Other considerations: part height support. Latest technology MTUs have features, a duplicate setup for alternate part supply, for enabling nonstop production.

Multi feeder Unit: MFU

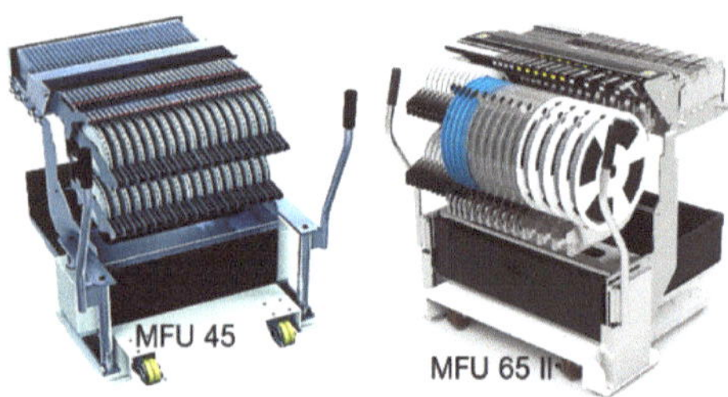

Allows the batch feeder replacement. External changeover can be performed. Reduces the change over time. Easy setting and removal. MFUs are dedicated to machine type, and as the name implies, MFU 65 has 65 feeder slots, and MFU 45 indicates the availability of 45 slots. With the help of MFUs, production changeovers can be performed in no time because it is an offline activity where operators load feeders with reels to get ready for a changeover.

Dip Flux Unit

This unit is used for the POP (Package on Package) application. Like the feeder, this unit is loaded onto the feeder pallet. Used to transfer the flux to bumped parts before placement. It supports varying bump sizes and high-speed heads. This unit contains a small squeegee that maintains the appropriate flux after each application. Flux is automatically supplied from the Syringe. Flux thickness can be adjusted. In some models, the dip tank moves back and forth to create an even film thickness using a squeegee.

SUMMARY

Innovations in parts supply technology reflected a lot on high mix low volume, high mix high volume production environments. To increase the efficiency of production, need to provide abundant feeder slots to perform multiple productions/with fewer changeovers.

Hexa feeder is one of the examples. It occupies three slots but accommodates six feeders. Specially designed for mobile manufacturing. Bulk component feeder technology or cassette-type part supply technology is under renovation to meet the challenges of Industry 4.0.

This technical content includes detailed descriptions of electronic manufacturing processes and the manufacturing enterprise that can help aspirants, students, and readers of these articles come to know "a little about everything." **Chosen Fuji as a role model in this context due to vast experience with Fuji Machines.**

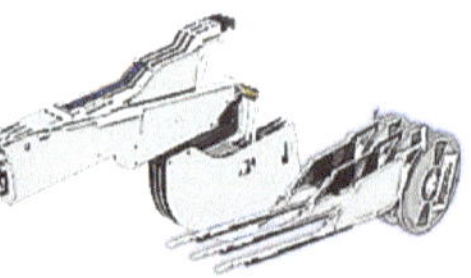

PLACING HEAD NOZZLES

Placing Head - NOZZLES

Quality nozzles and feeders are the core of pick-and-place. No SMT equipment can place components accurately and run efficiently without quality nozzles and feeders. These two factors are the core of the pick-and-place process. If the machine is either unable to pick parts consistently or hold on to the components during the transport from the feeder to the PCB, defects will result. An increase in defects means a decrease in production, costing the company more money over a short time. Proper feeder and nozzle and their maintenance are critical, especially with the current market growth and technology advancements in SMT equipment

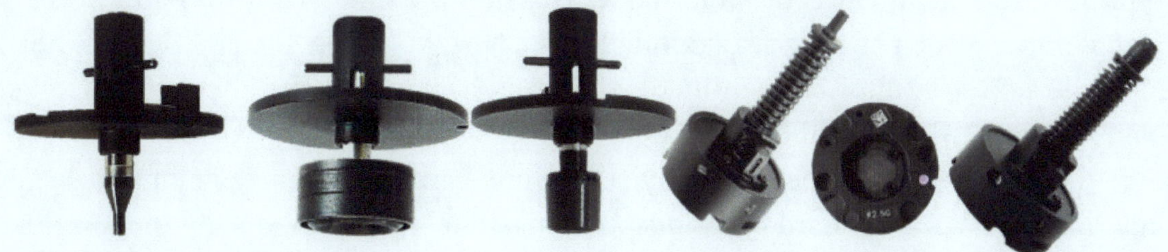

Nozzles are the first and last things to touch all components on a PCB. Pick-and-place nozzles can move tens of thousands of parts every hour (depending on CPH rating). These nozzles are required to hold the part firmly during transport while the placing head is moving and/or rotating at high speeds. With electronic component sizes reaching microscopic proportions, nozzle manufacturers must strive to maintain precision tolerances and exact dimensions in their designs.

Nozzles utilize the principle of vacuum suction to pick up the part from the feeder pick location and then place it on the assigned PCB location coordinates. While placing the component, the nozzle applies the principle of air blow to detach /release the part from the nozzle.

STANDARD NOZZLES

Standard nozzles are designed for most typical SMT applications to cover various components. Each vacuum nozzle is numbered according to the machine model, type of placement head, profile of the nozzle tip, and suction area. New generation pick-and-place heads offer automatic nozzle exchange without manual intervention. Nozzles are stored in compatible Nozzle Stations in the machine from which the placing head selects the nozzle as per the production program for the next component pick and placement process.

Typical nozzle hardware consists of a holding mechanism (locking pins), a disc with a reflective surface, and a nozzle tip (with negligible cushion movement). The structure of the Nozzles' geometry and holding mechanism varies with the type or model of the placing head. So dedicated nozzles for dedicated placing heads and dedicated nozzle stations as well.

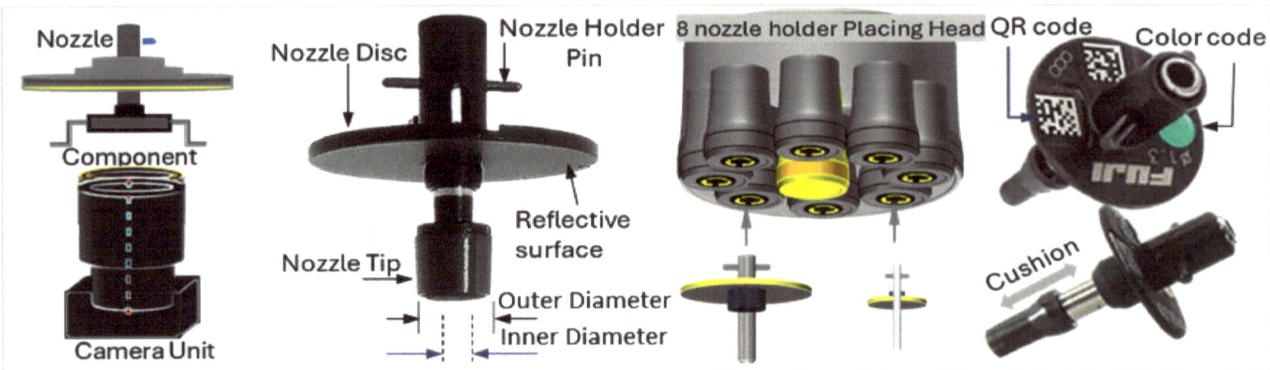

A typical nozzle has a QR code, color code, Nozzle tip diameter information in the legend printing, and brand. When nozzles are stationed in the Nozzle station, the Nozzle tip is not visible and the type of nozzle cannot be identified. Also, it is hard to classify the difference between 0.3mm and 0.4mm with a naked eye. So, operators can recognize the correct required nozzle with the help of color code. In the same way, a machine cannot recognize color codes. QR code plays a key role here. As a standard pick-and-place machine operation, during the nozzle change sequence, the mark camera (look-down camera) reads QR code data to identify the type of nozzles in each pocket of the nozzle station and their stored location.

Nozzle design is one of the most crucial ones, as the Placing Head's overall efficiency depends on nozzle performance. Nozzles are equipped with a reflective background, i.e., a disc with a reflective surface serves as the background for the Parts Camera, which ensures a good background during image capture by the machine part camera, allowing the target information to be highlighted smoothly during image processing.

Nozzle Selection

Nozzle selection depends on the type of component picked up and the place of the circuit board. The selection of the Nozzle number depends on component shape and weight. Though nozzle selection is predefined in the machine library, sometimes manual intervention is required for odd-form component packages.

Outer Diameter	Applicable Parts	Outer Diameter	Applicable Parts
Ø 0.2 mm	0201 (008004)	Ø 2.5 mm	SO IC 8pin
Ø 0.3 mm	0402 (01005)	Ø 3.7 mm	SOIC 16pin
Ø 0.4 mm	0603 (0201)	Ø 5.0 mm	SOIC 28pin
Ø 0.7 mm	1005 (0402)	Ø 1.3 mm MELF	Melf
Ø 1.0 mm	1608 (0603)	Ø 1.8 mm MELF	Melf
Ø 1.3 mm	2012 (0805)	Ø 2.5mm MELF	Melf
Ø 1.8 mm	3216/3225, (1206/1210)		

Custom Nozzles

Custom nozzles are Unique Nozzles specific to the odd-form component. Odd-form components are those that cannot be easily handled by standard nozzles because of odd-form shapes and irregular sizes. Hence, special nozzles need to be developed for smooth pickup and placement.

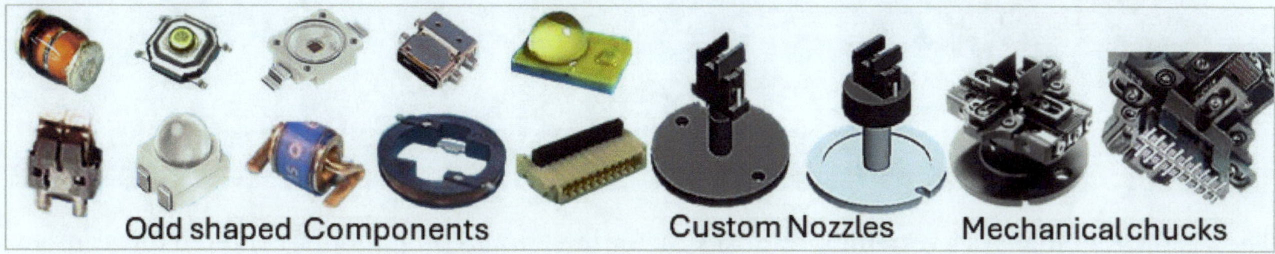

Custom nozzles and tools, e.g. mechanical chucks, nozzle grippers, etc., normally developed by placement machine manufacturers - OEM range of machines - as per user request, as these nozzles are very specific to the type of placement machine and head configuration. The design of a custom nozzle varies from component to component and component package.

The vision system plays a very important key factor in various components, centered on accuracy and repeatability. Capturing the perfect image of the component is very much essential for a vision system. An industrial machine vision system means Accuracy, repeatability, Low cost, robustness, and high mechanical and temperature stability. The design of the lighting system, the Lens structure, the camera type, and the Image processing systems' speed determine the machine's capability, accuracy, and repeatability. To achieve the required results in the vision system, part supply location tools must have enough intelligence to compare the trained pattern quickly and accurately to the actual component image.

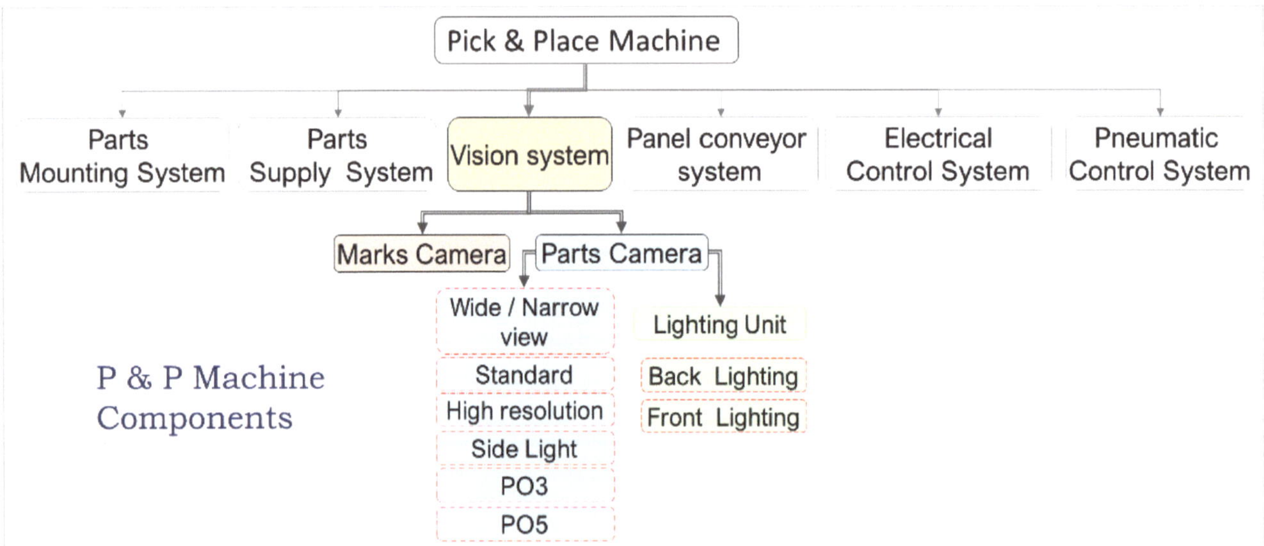

Pick-and-place machines utilize machine vision applications for guidance, identification, gauging, and Inspection. These applications provide location information obtained by pattern matching (with the help of vision software tools), position and orientation of the part, inspection, and measurement. They then compare measured data to specific tolerances and ensure the correct angle (machine to align the part) to verify proper assembly.

Hence, machine vision systems rely on digital sensors fitted inside industrial cameras with specialized optics to acquire images so that computer hardware and software can process, analyze, and measure various characteristics for decision-making. There are four types of methods for centering component pickup and placement.

1. No Centering mechanism

In this case, the part is picked up by the tool/gripper, called the pick-and-place head tool, from the specified location or coordinates and placed on the PCB at the specified location without any corrections. So, if the part is picked off-center on the part supply, it will be off-center when placed on the board. This is not useful for a precision production environment.

2. Mechanical Centering (Using the Jaws)

In this method, the component is picked up from the supply and moved into its center position in the X and Y axes on the pick-up head. Repeatability can be expected within +/-.001" accuracy. The mechanical centering concept is no longer used in present technology.

3. Laser Centering

In this method, the component is picked in line with a laser beam, which detects the component's center position on the tool head and recalculates the zero point of the part according to its position in the X, and Y axes and rotational position relative to the head for an accurate placement on the board. Less reliable and limited on the type of part handling. Cannot center parts below 0402 packages or larger than 35 mm square.

4. Vision Centering:

Two types - "Look - Down" and "Look - Up"

1. Look–Down Vision Centering

Look-down vision will view the top of the component before picking it up for its pick-up location. It compares its viewed image file from the stored database, then calculates the center of the component, adjusts the pickup tool position coordinates as per calculated offsets, picks up the component, and transports it to its programmed location on the PCB board for placement.

It is True touchless centering, can handle odd-shaped and delicate components, and is accurate to +/-.004". However, this method of centering is slower due to the time slice required for processing. For Look-Down vision, the part may move from its pickup point to its placement on the board. The capable range is 0402 – 15 mm. Therefore, the present technology does not adopt this method.

- Look-Up Vision method

This type of vision method is the most influenced by present technology. It is the most accurate centering method. Here, the target component blindly picks up from the pickup location without pick-up inspection and moved to a Part camera station that looks at the bottom of the component, and calculates its center position based on the Vision system's vision processing database that identifies the image geometry, stored in the machine's database. A slower method of centering due to processing time.

The Parts camera is generally physically fixed at a specific location in the machine work area. With the help of a vacuum tool or gripper, the machine head system picks the target component from the parts supply system and moves the picked component above the Parts camera for inspection. To capture the component's bottom image, the parts camera looks up and carries image processing in all aspects of geometry. Based on image results, the vision system communicates with the Machine Head system for corrective action and performs component placement on the PCB accurately and precisely.

COMPONENT PLACEMENT PROCESS IDEAL CONDITIONS

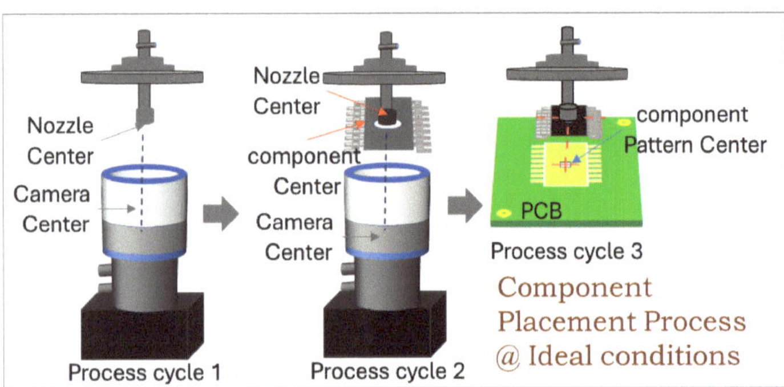

Component Placement Process @ Ideal conditions

Let us consider the ideal condition. As shown in the figure, the parts camera center and vacuum nozzle or gripper tool center are aligned and calibrated as a process 1 to carry out further processes. The vacuum nozzle picks the component from the center of the component at the parts supply location. The nozzle and the picked component move above the Parts camera at the predefined location at which the nozzle center and camera center are straight. Since the nozzle picks the component at its center, the nozzle center, component center, and camera center are aligned and straight. Hence, there is no offset in the XY coordinates. Further, the component is moved to the placement position on the PCB, and the component is placed precisely on the component pattern.

COMPONENT PLACEMENT PROCESS @ PRODUCTION MODE

Now, consider the component placement process in production mode. Here, the offset coordinates for pickup and placement need to be considered for each pickup and placement of a component. As shown in the figure, both the parts camera center and vacuum nozzle or gripper tool center are aligned and calibrated as a default process for all situations like nozzle change, head change, or machine reboot.

In this scenario, the vacuum nozzle picks the component from the parts supply location in production mode. Hence, there is no guarantee that the nozzle picks the component without inspection at the parts supply location. Let us consider the component not picked from its center, rather a little offset. There is a pick-up offset in coordinates from the target center coordinates. In

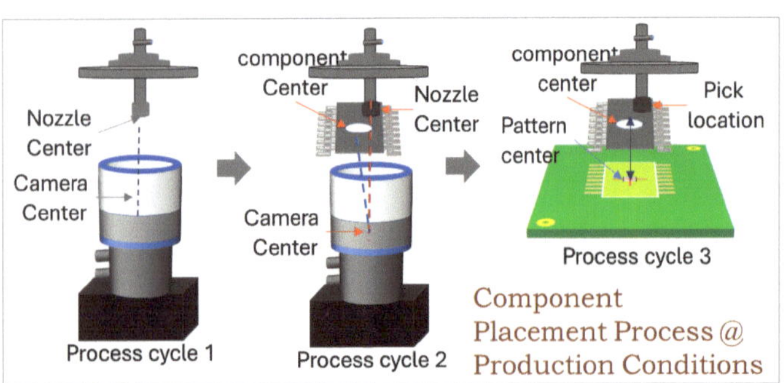

Component Placement Process @ Production Conditions

the next cycle, as a default process step, the nozzle, along with the picked component, moves above the Parts camera at the predefined location at which the nozzle center and camera center are straight and calibrated.

The look-up Parts camera captures the component image, checks component geometry and algorithms as per the stored data, and checks the processed image center of gravity and orientation of the picked target component. Pick-up offsets and orientation were calculated with reference to the camera center, and measured offset information was processed to the Central Processing Unit CPU so that placement angle and offset corrections take place by the Head motors. Further, the component is moved to the placement position on the PCB and placed precisely on the component pattern.

MARK CAMERA

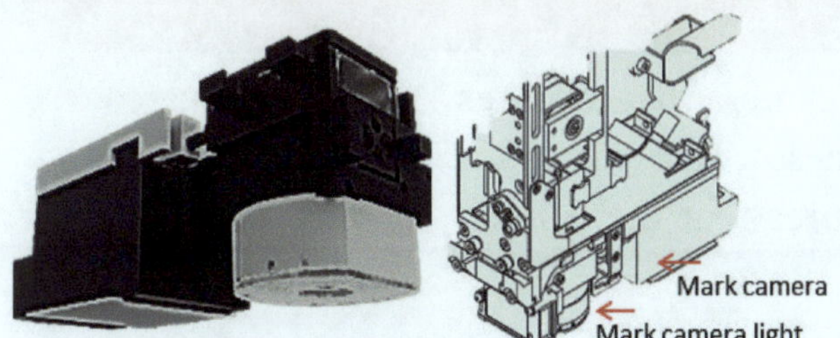

Fiducial or Mark Camera

In the pick and place placement process, there is no guarantee that each time the PCB moves in the conveyor and stops exactly at that X, Y location. Offsets may occur due to the mechanical movement of conveyors. Further, there may be acceptable tolerances in PCB dimensions and Fiducial locations when supplied by the supplier. However, these tolerances cannot be granted for component placement offsets. Hence, in placement machines, the mark camera is mounted with the movable Placement head so that vision happens from Top to bottom. For every throughput, the mark camera measures the actual coordinates (X and Y) of fiducial marks on PCB/Panels being produced so that coordinates can be corrected in the placement program for that panel.

Fiducial cameras in the latest machines are utilized for many applications - used to look for Global fiducials and local fiducials on PCB, component tracing after placement, to read fiducials on feeders, Nozzle stations, and clamping on conveyors for position alignment inspection. Also used to read QR codes on Nozzle stations and Nozzles to recognize the model/type like many other applications - inspects part empty in reels or trays.

PARTS CAMERA

Parts camera, often fixed and mounted in the machine base so that vision happens from bottom to top, captures the image/shape of parts as they pass through the camera's field of view. The system uses software to calculate the distance between various points in the captured image or geometric location on an object and determines whether these measurements meet specifications. If not, the machine vision system sends the failure signal to the machine controls and triggers a reject/ corrective action mechanism. Many machine vision systems can measure object features within 0.0254mm.

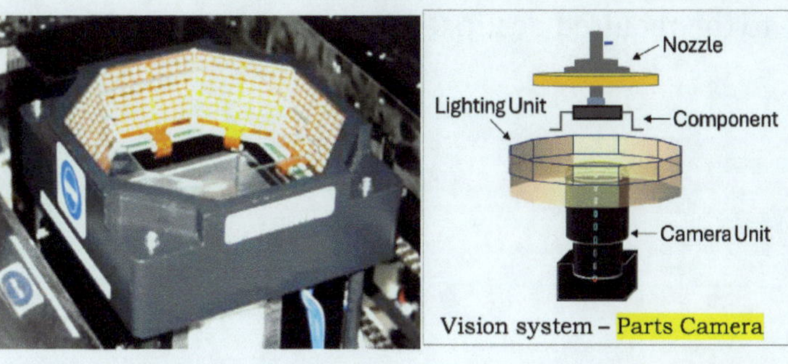

The part camera sees that the part is not exactly centered on the tip and has some unwanted rotation. The machine rotates the tip until the up-camera sees that the part is in the "correct" rotation. Then, the machine uses the up-camera to figure out what X and Y compensation it needs to use to get the part to the correct location on the board.

The vision system also offers important tasks such as Camera calibration, Matrix jig measurements, and assembly support functions such as Placement accuracy tests, Nozzle center measurements, PCB position offsetting, and Component confirmation.

HOW DOES VISION WORK?

Vision-enabled pick-and-place head systems have a camera that captures images of components as they travel. The system software identifies and processes these images. The lighting unit ensures that the camera captures the best possible image.

Vision-enabled pick-and-place machines see the position of a picked part and adjust themselves accordingly to manipulate it. The main objective of this work is to develop a system to see the component position, calculate the component's position, grasp it, inspect it, and move it to where it needs to be placed on the PCB.

The camera in a robot vision may be defined as a process of extracting, analyzing, and understanding the information from the images. In this work, we use a CCD (charge-coupled device) camera to capture two-dimensional images from the target object. CCD camera discretizes the scene into a 752 * 552 format, which can be stored in the temporary storage RAM of the computer.

- Vision System Functions
- Presence Inspection
- Binary Processing
- Blob Analysis
- Practical Positioning
- M/c Vision Identification
- Flow detection
- Detection Through Measurement

The information from the CCD camera is processed by an image processing card mounted within the computer. Automated Inspection software is used for processing the image (geometric matching, Shape, and color analysis), and to find out the center of the object and its orientation. The view area is digitized so that the coordinate values of any object in this area are found.

Communication triggers and inspection results are directed to industrial devices over serial and Ethernet protocols. It also includes the ability to set up complex pass/fail decisions to control digital I/O devices and to communicate with serial devices. An interfacing card is used for connecting the computer to the robot. Values are sent to the pick-and-place head system via an I/O card, which makes the machine placement head position its gripper accordingly to pick up the object and move it to the required destination.

COMPONENTS OF MACHINE VISION SYSTEM

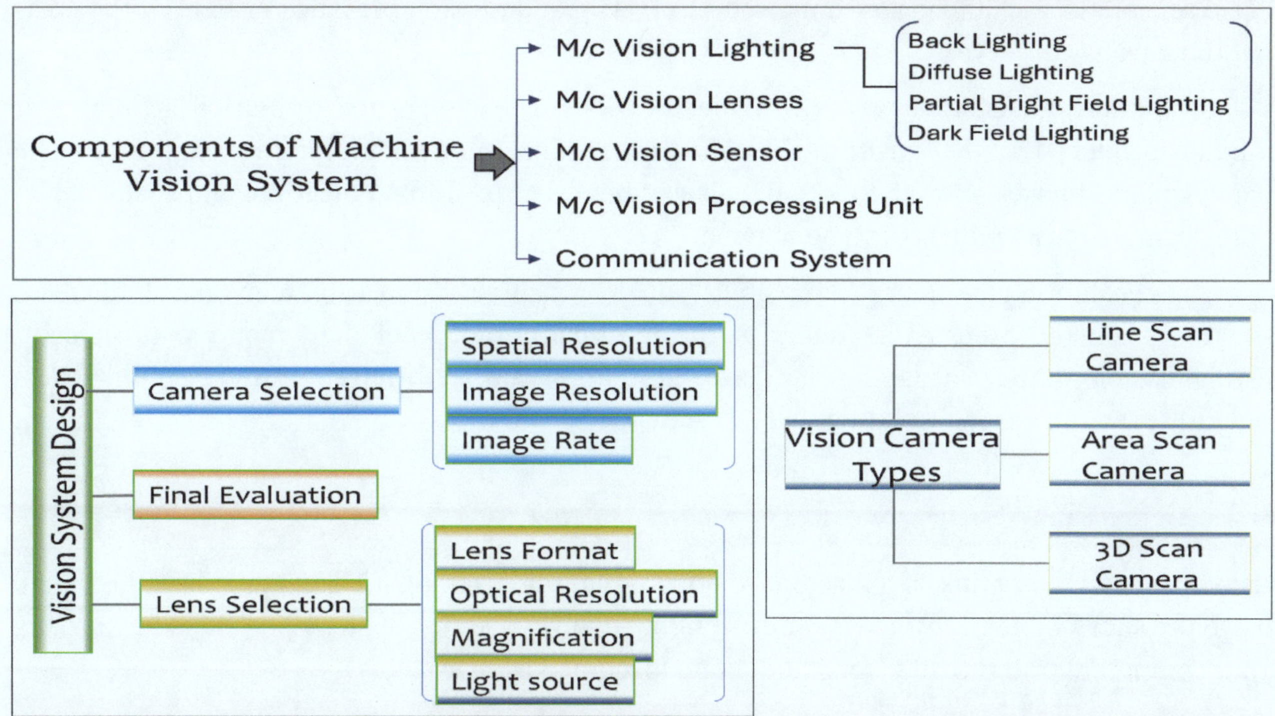

MACHINE VISION LENS

- Lenses characterized by Field of View, Depth of Field, Depth of Focus, and Aperture, which describe the image quality they can capture.

- Field of view refers to how much area the image sensor views; lenses with higher focal lengths have a reduced field of view.

- Depth of field refers to the ability to maintain acceptable image quality without refocusing if the object is moved farther from the plane of best focus. It also influences the object's range of acceptable motion.

- Depth of focus refers to how the quality of focus changes as the sensor is moved while the object remains in the same position.

- Aperture is the opening of the lens through which light passes to enter the camera. It controls the amount of light entering the lens. It is inversely related to the depth of field. (Old technology machine vision systems used different methods to obtain different frame types. For example, Camera 7 used a wide view and a narrow view, which are obsolete technology at present. Two cameras of the same type are mounted one lower in the machine to provide a wide view and one higher in the machine so the field of view is smaller but resolution is higher.)

IMAGE SENSOR

The image sensor inside the machine vision camera converts light captured by the lens into a digital image. It typically utilizes a charge-coupled device (CCD) or complementary metal-oxide-semiconductor (CMOS) technology to translate photons into electrical signals. The output of image sensors is a digital image composed of pixels that show the presence of light in the areas that the lens has observed.

Resolution and sensitivity are critical specifications of image sensors. Resolution is the number of pixels produced by the sensor in the digital image. Sensors with a higher resolution produce higher-quality images, meaning more details can be observed in the object being inspected, and more accurate measurements can be obtained.

The resolution also refers to the ability of the machine vision to perceive small changes. Sensitivity, on the other hand, refers to the minimum amount of light required to detect a distinguishable output change in the image. Resolution and sensitivity are inversely related to each other; an increased resolution will decrease the sensitivity.

VISION PROCESSING UNIT

Vision processing is a unit of a machine vision system that uses algorithms to analyze the digital image produced by the sensor. Vision processing involves a series of steps performed externally (by a computer) or internally (for stand-alone machine vision systems).

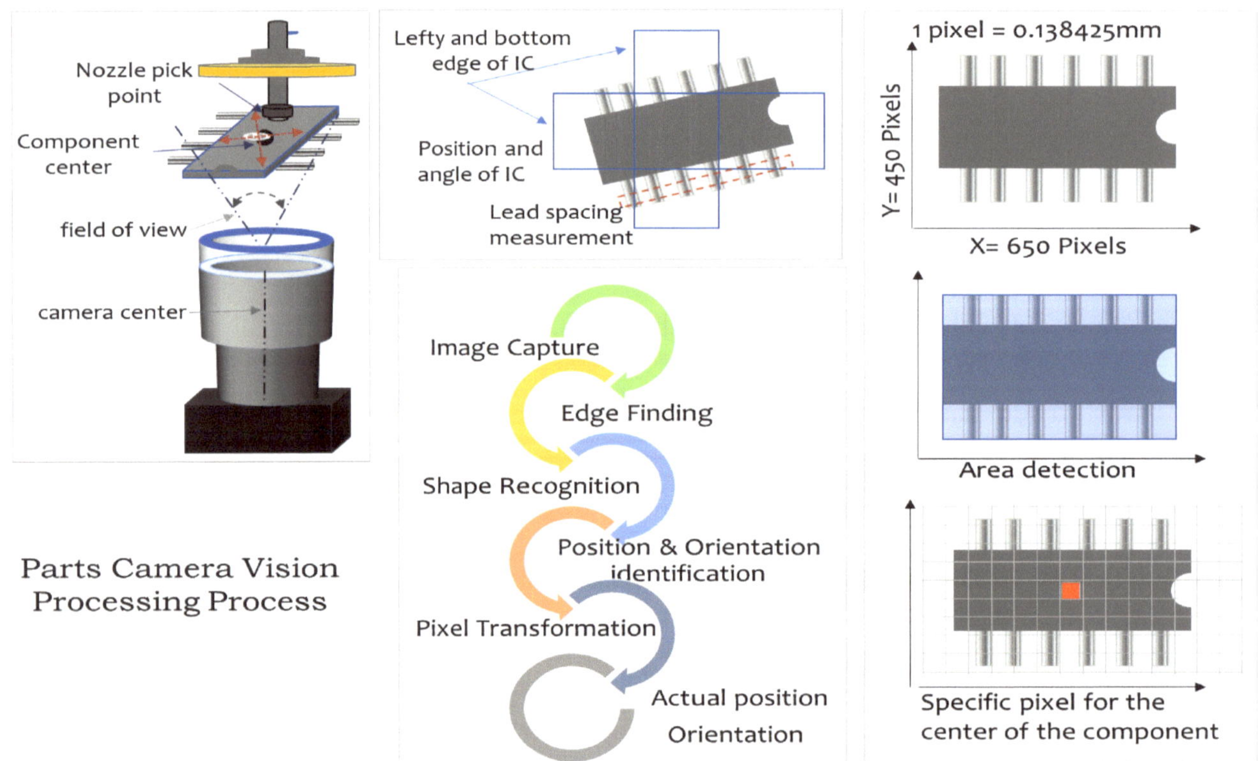

First, the digital image is acquired from the image sensor and is relayed to the computer. Next, the digital image is prepared for analysis by making the necessary features of the image stand out. This image is then analyzed to locate the specific features that need to be observed and measured. Once observations and measurements of the features are completed, they are compared to the defined and pre-programmed specifications and criteria. Finally, the decision is made, and the results are communicated with the Input/output interface.

If the captured image is done correctly, a sharp contrast is required between the background and the target object. Because sharp contrasts are easy to recognize, the shape is easy to analyze. I.e., capturing image and image analysis are directly related.

The machine vision system first acquires the image from the Camera using vision acquisition software, communicates with a set of camera hardware drivers, and displays the image on a monitor, the image type, and its coordinates.

The captured image, i.e., the digital value provided by each element, is transferred to the image processor for further processing of pattern-matching algorithms(threshold) and edge detection algorithms.

A pattern matching algorithm is used to search for a pattern or template image in an inspection image with the template image (Shape data) saved in the computer to check whether both images match or not.

Edge detection algorithm used to find the object dimension and used to display the object edges to find the objects' X and Y pixel values. Uses the concept of the difference between the light and dark areas of an object to detect the edges. Compares it to the image stored in the processor's memory.

Threshold- This is the binary conversion technique in which each pixel is converted into a binary value, either black or white.

VISION LIGHTING

Lightning is the one key to success in machine vision results. The complexity of the image-processing algorithm depends on illumination. A target object in our application is an electronic device seen by the camera. Proper lighting should provide contrast and minimize reflections and shadows. The lighting technique involves the Light source and its placement concerning the part and the Camera. Hence, image enhancement purely depends on the design of a particular light technique. One thing that needs to be understood is that a machine vision system creates images by analyzing the reflected light from the object, not by analyzing the object itself.

LIGHT SOURCE METHODS

There are two methods of lighting available for the part cameras. The ring lights mounted on each placing head comprise the backlight system. The ring lights mounted just above the part cameras make up the front lighting system.

Lighting sources commonly used in machine vision systems are Fluorescent, Quartz halogen, and the most widely used LED source. The choice of the lightning source depends on the

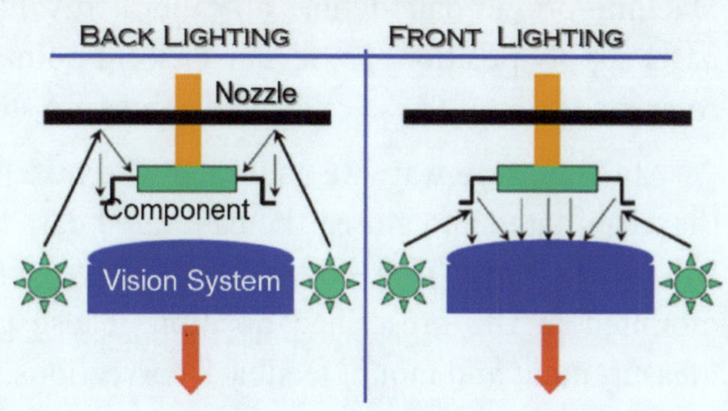

application. LED light sources are widely used nowadays because of their longer life expectations, stability, and intensity. Quartz Halogen can be the choice for a particular inspection because it offers greater intensity.

COMMUNICATION SYSTEM

The communication system quickly passes the decision made by the vision processing unit to specific machine elements. Once the machine elements have received the information (or signal), the machine elements will intervene and control the process based on the output of the vision processing unit. This mechanism is accomplished by discrete I/O signals or data communication by a serial connection in the form of RS-232 serial output or Ethernet.

IMAGE SENSOR

The lens captures the image and presents it to the sensor (CCD array used as a sensor) in the form of light. The sensor in the Machine vision camera converts this light into a digital image, and this data is sent to the processor for analysis.

CCD and Metal Oxide Semiconductor (MOS) cameras can collect and store charges. Made up of a matrix of CCD elements referred to as a pixel. The common field of vision to capture the component's image. For example, adjustable CCD arrays range from 512 x 512 up to line sensor camera 2048 pixels. Each element provides a digital value that represents the intensity of the light incident at that point. A value of 0 represents black (no light), and a value of 255 represents white. All values in between represent varying values of the grayscale.

Image sensing Cameras like Standard type, High-resolution type, P03, Standard P04CL, wide view P05, P05CL, etc., are the most common in present Machine vision system technology with lightning options, light, combination front light, and side light. If you consider the P05 camera as an example, it is a high-speed camera with a Field of view of 75 x 75 mm and resolution 37 um/pixel. Recognizes minimum pitch: 0.3/0.4(lead/ bump) and minimum width/diameter: 0.15/0.25 mm (lead/bump) with minimum interval: 0.15/0.15 mm (lead/bump) Vision processing

VISION CALIBRATION

Machine-proper data defines all calibrated values for the position of the Fixed Parts camera and parts supply position. So, feeder Pick-up points and tray pick-up points are fixed with machine reference points. These calibrated values are stored in the proper data.

In the same way, we calibrated the nozzle center point and the camera focus points, and then calibrated and stored the calibrated data in the proper data. The latest machines for stable and measurement corrections, a special type of Gauge chip parts provided by the manufacturer, mounted in the prescribed position. These gauge parts are used for nozzle rotation center measurement and mount feedback corrections.

The machine robot picks the component concerning calibrated pick positions and brings the component to the calibrated camera focus point so that offsets between the component center and camera center can be calculated for placement corrective action.

EXAMPLES -CCD CAMERA PARAMETERS

Vision Size - When using a Rear camera, the user must set the height at which the parts are to be vision processed.

CCD Level –Offset - This item allows the user to inspect a part using a different CCD level from that set on the machine. The CCD level determines the brightness of the camera during vision processing. When the image of the part is too dark relative to the image of the background, input a larger value to make the contrast clearer. A larger value makes the image clearer, and a smaller value makes the image less clear.

Camera Type - This specifies the field of vision of the part camera. The size of the part processed determines the setting of this parameter.

CONCLUSION

Despite powerful offline software helping production programmers to get the best optimization results in terms of feeder arrangement and other factors, the thirst for the best cycle time never stopped. Design engineers made changes in the Machine layout and even peripherals. Machines were updated with high-speed, high-resolution machine vision cameras. In some models, the Parts camera is fixed to the Robot itself so that picked components can be inspected while traveling, i.e., during the placement. This feature is called Fly on Vision. Parts camera fixed in placing head itself to make fly on vision. Fixed Part cameras on the Machine Base were made very close to the conveyor at an appropriate location to reduce traveling time. Similarly, Feeders are made to compact their feed positions further inside the machine so that pick-up points become very close to the PCB work area.

The main objective of the pick and place machine production program is to pick the required component from the component center, from the defined location, with the defined nozzle, inspect the picked part with the help of the Vision system, and place that component on the desired location on the PCB precisely with offset corrections. Ideally, perfect placement happens only when coordinates match the center of the element (here, the target component) and the center of its footprint on the PCB with proper orientation and placement height.

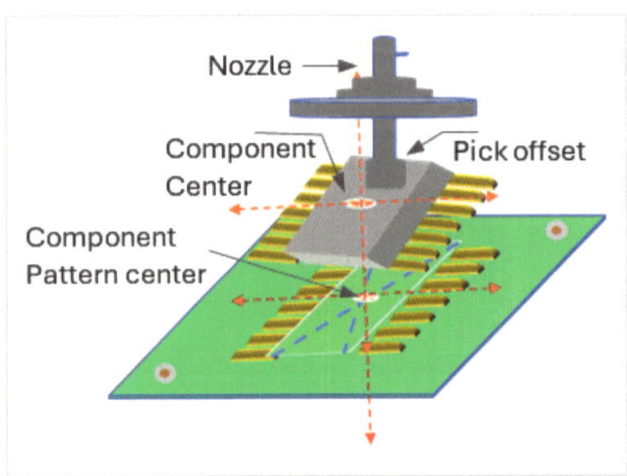

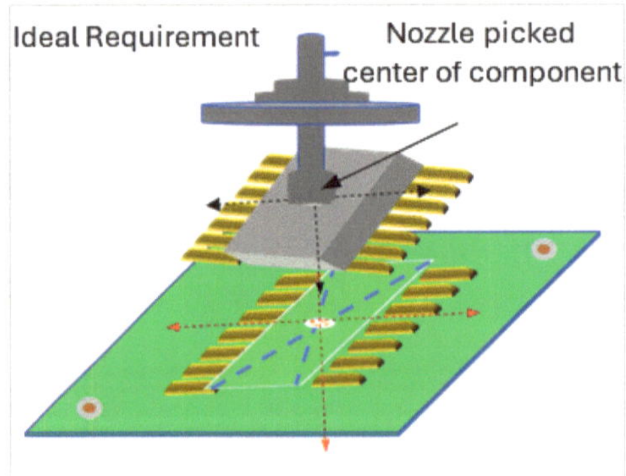

The objectives can be achieved with the Centroid data of all footprints where the components will be placed on the PCB. Similarly, we should have centroid data for every element to be placed. The component manufacturer provides this data in the form of a datasheet.

The data sheet provides Part information, such as the centroid, part height, lead information, and other geometrics required for part inspection and pick-up offsets. This data is stored in a centralized server location as a part of a library with a unique identity for each electronic package.

Similarly, PCB data provides information like Pad information of all component footprints on the PCB surface, Centroid coordinates of component footprints, and other geometrics, which are required for Precision placement and offset calculations. This data is stored in a centralized server location as files with a unique PCB identity. Machine vision systems retrieve this data during every pick-and-place process and make decisions about that placement.

Machine manufacturers develop sophisticated and highly user-friendly offline software so that programmers can create realistic placement programs without stopping the machines in production lines. Offline software is the most convenient method for programming the pick and place for creating new jobs for PCB assembly without interrupting production.

All functions, such as editing the programs, creating new programs, simulation, creating shape or part data for odd-form components, and monitoring production performance, can be done offline without disturbing the production lines. Options like importing and exporting part data and shape data library from other sources or from the data sheets, which make programming easy and fast operation.

Many OEMs like Siemens, Panasonic, Fuji, and other similar associates develop SMT programming software that helps in the programming of pick-and-place machines for the production of Electronic PCB assemblies. Fuji Flexa and Nexim are well-known and very popular offline software from Fuji pick and place equipment manufacturers for programming the pick and place machines. **In this context, Fuji Flexa offline software has been demonstrated for understanding purposes, wherever applicable, so that readers can understand.**

INTRODUCTION

For typical programming, the machine requires Centroid data, Bill of material data (BOM), Printed circuit board data, part data, and mark data. The programmer must have this data and set up data, as the programmer should know about the machine configuration and resource availability (i.e., feeders, nozzles, Placement heads, tools, etc.) so that the job will be created efficiently and production can happen effectively.

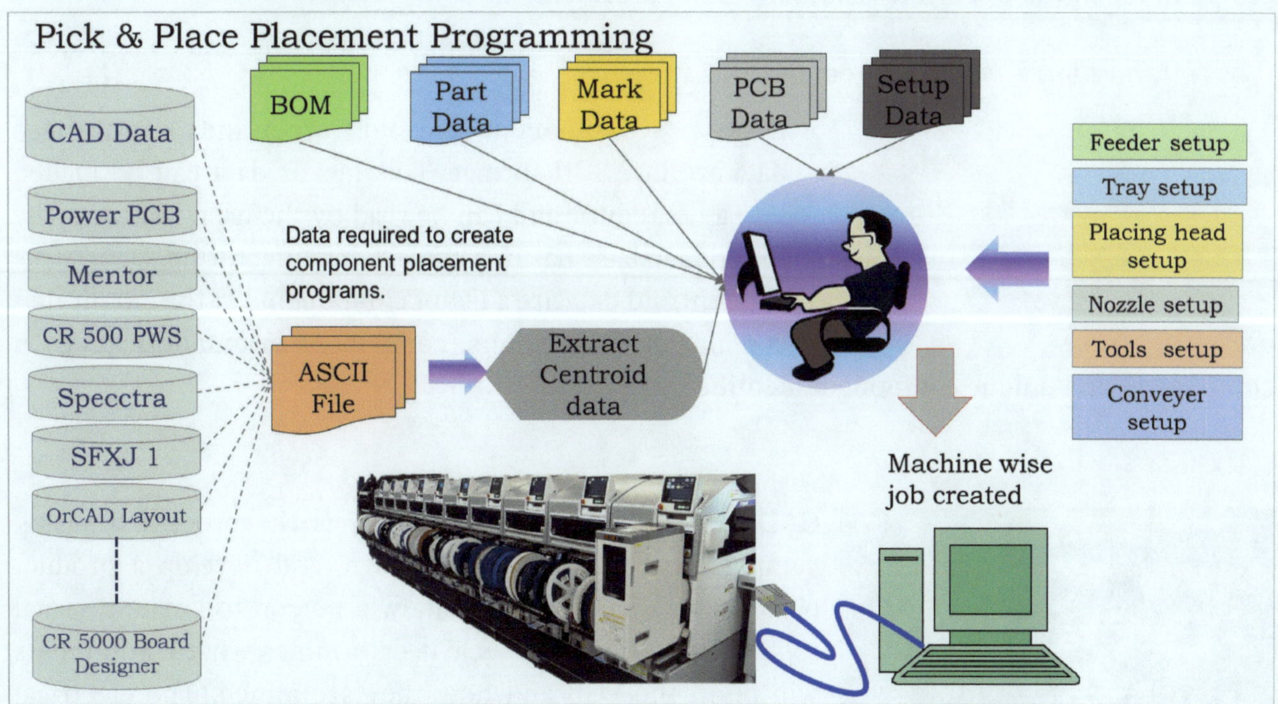

CAD DATA: COMPUTER-AIDED DESIGN DATA

CAD information plays a critical role for the product Programmers. Higher component pin counts and smaller part footprints make it essential to place components with greater accuracy. So, manufacturers use CAD Data to seek ways to develop manufacturing programs and documentation faster, cut manufacturing cycle times, reduce errors, and improve manufacturing quality. Eliminates manually defined placement information with a digitizer or vision camera. Also eliminates human errors when generating the data.

Product assembly programmers obtain product (PCB) CAD data either from the PCB designer or from the PCB supplier. CAD data contains all the information needed to help manage the manufacturing process. Using the original CAD system, the programmer gets the exact component coordination system, i.e., centroid data– absolute accurate information that specifies the position at which parts are to be mounted.

CAD also provides information on the PCB Layer information, Test point information, and coordinate information about Pads, Via holes, and Centroid (X, Y and), component reference. But for assembly programming, only centroid data is mandatory. The same CAD data is used for PCB manufacturers and stencil Manufacturers.

Normally, the assembly programmer gets centroid data from the source, but sometimes this data can be created by various CAD forms, developed by various manufacturers such as Mentor, Power PCB, CR500, etc. Hence, the programmer may not get the same type of CAD Data from the suppliers to make the product job for his production line placement equipment. In such cases, the programmer extracts centroid data from CAD ASCII text.

What information is available from the CAD system?

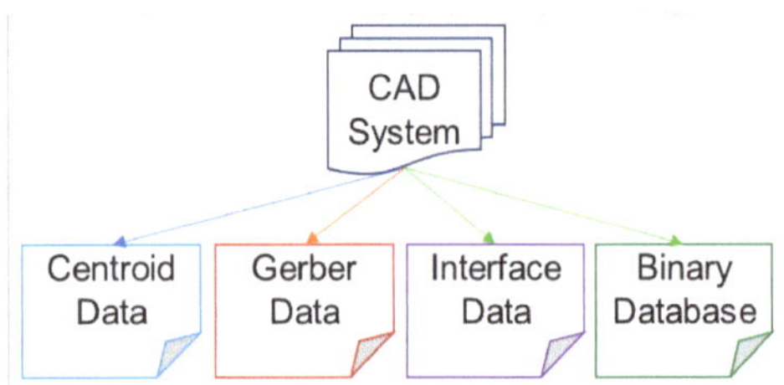

Data types- Component Centroid, Gerber, and CAD interface data are in ASCII format. This means data can be loaded into a text editor and can be read by the user. Whereas the Binary database has no value for manufacturing. Data types for Centroid data are a list of components in the circuit that contains X, Y locations, component orientation for each component, and unique component identifier.

CAD Data

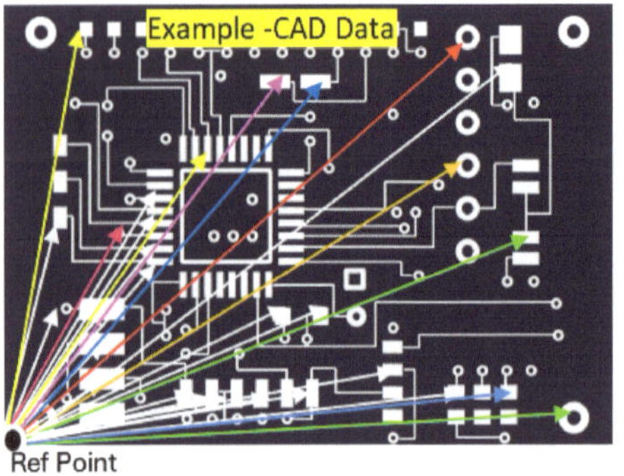

CAD, or Computer Aided Design, is the process of using computer software to draw, design, and develop a product or concept. CAD programs allow a person to create a visual representation of an object. It determines what components will be in the PCB and how they are joined. The electrical traces on a schematic are called 'nets'. A designer can use an existing schematic design or create their own. A designer will run simulations to test how well the schematic will operate as an actual circuit. CAD files include the DFM settings used for your board layout, as well as the component placement information necessary for the PCB to be assembled.

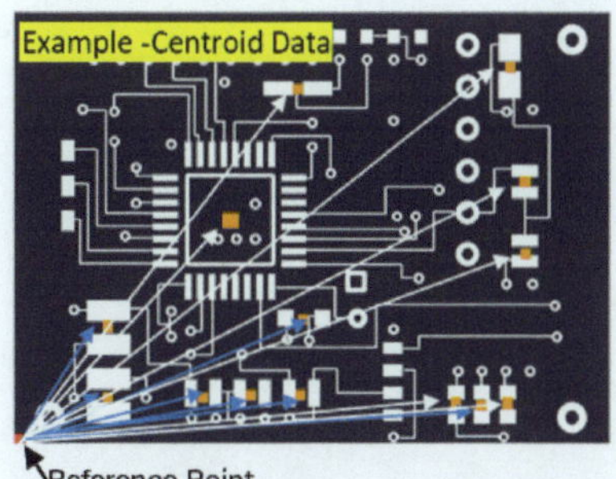

Component Centroid Data

Centroid data must be required for the component placement programming. It is a centroid-formatted text file containing part placement coordinates, X Y Location, orientation, and unique component identifier. Offline programming software requires this data in text format.

RefDes	Layer	LocationX	LocationY	Rotation
C2	Bottom	3.03	1.08	0
IC2	Top	3.25	0.875	180
IC3	Bottom	3.25	0.9	0
R2	Top	3.425	0.8	90
R3	Top	3.545	0.105	270
R4	Bottom	1.475	0.5	270

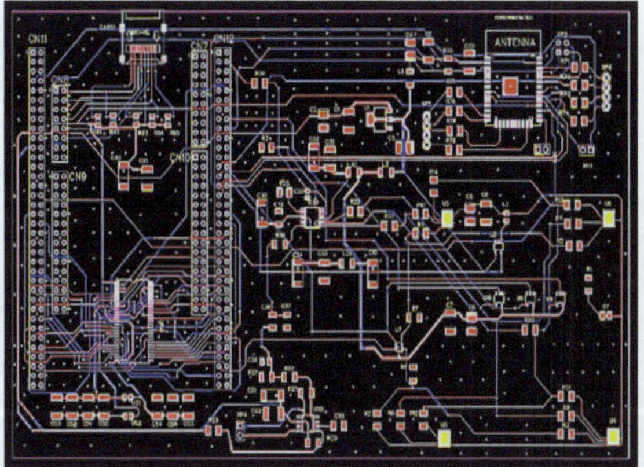

Gerber Data

Gerber data is data output by a CAD system, widely used by PCB designers for creating the film artworks used to fabricate the circuit. It's a simple 2D (ASCII files) artwork Image file of fabrication data includes drill data (vias and holes), pads, drill files, and board profiling files like Paste Mask layer, copper layers, silk screen layers (marking on PCB), and solder mask layers (solder resist) defined by series of vector coordinates. Gerber data represents a picture of the circuit well suited for making artwork films, especially for SMT stencils.

CAD Interface Data

Interface data contains board shape, fiducials, test pads, traces, and component geometry information. This is a unique file format for each product. Various software applications allowed reading this information for product programming needed for assembly.

BOM: Bill of Material

A Bill of Material is a list of the raw materials required to make that product, till the end of the production process, including packing. It means this Excel file contains electronic parts, surface mount devices, through-hole components, Electrical parts, mechanical parts, subassemblies, harness parts, tags, stickers, packing material, and even manuals, which cost to a Product. BOM also refers to part number, part description, specification, quantity, and reference locations that are needed to manufacture a product.

The programmer needs to extract or segregate BOM data as per the Machine capability for which the Component needs to be placed. The extracted BOM data is prepared normally in digital format, preferably an Excel sheet, containing the following information.

Item	Qty	Reference	Value	Description	MFG	M P N	PCB Footprint
1	1	C1	4.7uF	CAP CER 4.7uF 10V X5R 0402	MURATA	GRM155R61A 475MEAAD	C0402
2	8	C2,C3,C75,C77, C81,C82,C98,C99	22uF/10V	CAP CER 22uF 10V X5R,20%, 0603	Samsung	CL10A226MP CNUBE	C0603
4	5	C5,C60,C61,C93, C102	0.01uF	CAP CER 0.01uF 16V X5R 0402	Yageo	CC0402KRX5 R8BB103	c0402_top
5	10	C7,C12,C13,C16, C19,C22,C43,C57 ,C58,C84	1uF	CAP CER 1uF 16V X5R 0402	Yageo	CC0402KRX5 R7BB105	c0402_top
6	5	C28,C76,C85,C92 ,C100	10uF/6.3V	CAP-CER-10uF- X5R-6.3V-20%- 0402-SMD	Murata	GRM155R60J 106ME15D	C0402
7	1	C31	15pF	CAP CER 15pF NPO 50V 5% 0402	Murata	GRM1555C1H 150JA01	c0402_top
8	1	C32	10pF	CAP CER 10pF NPO 50V 5% 0402	Murata	GRM1555C1H 100JA01	c0402_top
9	8	C38,C41,C45,C47 ,C50,C62,C96, C97	10uF	CAP CER 10UF 16V X5R 0603	Yageo	CC0603MRX5 R7BB106	C0603

Position (reference designator)
- ☞ Quantity
- ☞ Description of component
- ☞ Value
- ☞ Material requirement
- ☞ Power Rating
- ☞ Tolerances
- ☞ Package or case type
- ☞ Supplier Name

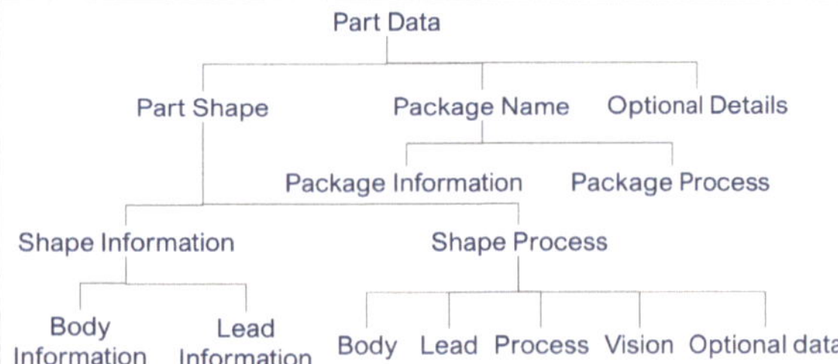

Part Data

Part data describes and provides the data of every type of component that needs to be picked, undergo the vision process, and placed on the circuit board. Part data for a specific component can be acquired from the database, i.e., part data library, or new data can be created. Part data provides the details – Shape data and package data.

Shape Data

Shape information provides the details of the part's body size, i.e., length, width, height, tolerances, and body color. The shape process provides the details of lead information, the Vision type number recommended for the part, the Nozzle to be used for that part, and certain other parameters that apply to those specific parts.

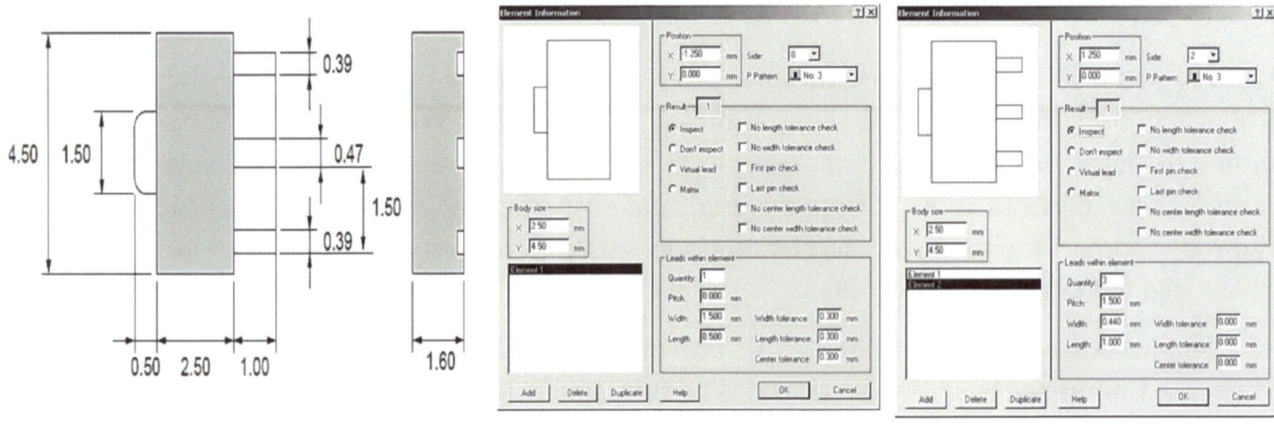

E.g. Vision 6 Part shapes

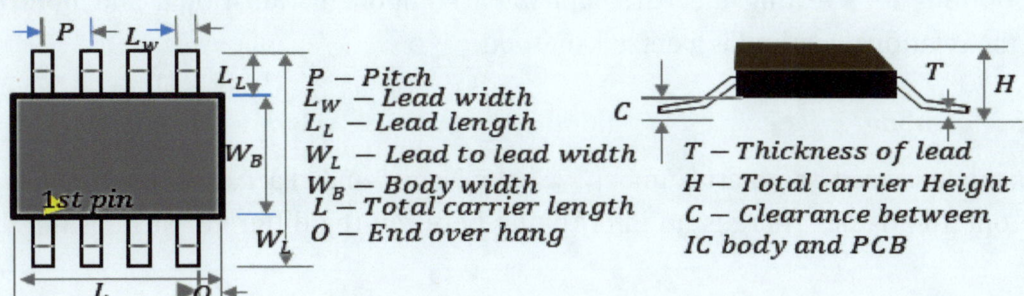

Vision Data

The vision processing methods are classified depending on the part shapes, the machine types, and the type of parts camera, and each vision processing method (algorithm) is assigned a number.

Vision type specifies a vision type number to select the algorithm to be used when inspecting parts. Normally, the equipment manufacturer provides this information. Similarly, we need to provide details of offsets during image capture and exposure time between multiple images.

Direction

This item specifies the direction of the parts in the package. It determines the part direction when viewing the sprocket holes on the left.

Side Data

To distinguish between the four sides of the part, each side was assigned a number.

Lead Information

Data entry here is necessary for parts with leads. Input a "0" or a "1" to indicate which positions on each side of the part have leads. Input "1" for positions where leads are present and "0" for positions without leads.

Element Data

When defining lead configuration for parts with non-uniform leads, divide the leads into groups within which all leads are uniform in length, width, and pitch. Each group is termed a "lead element". To distinguish between lead elements, each lead element is assigned a number. When assigning numbers to lead elements, use the same orientation used for inputting Side data. Beginning with the lead element on Side 1 (the lower side) nearest the lower left corner of the body, the user counts the elements, numbering them Element 1, Element 2, etc.

Shape Process

P — Pitch
L_W — Lead width
L_L — Lead length
W_L — Lead to lead width
W_B — Body width
L — Total carrier length
O — End over hang
T — Thickness of lead
H — Total carrier Height
C — Clearance between IC body and PCB

Provides the details of lead information such as Lead Length X, Lead Y, Lead pitch, and Lead width. Also provides the details of an applicable nozzle to be used, the Vision type recommended for the part, and certain other parameters and special marks (first pin, orientation) which apply to those specific parts.

Package Data

The supplier supplies parts in different packages – Reels, sticks, bulk, or trays. When creating a placement program, the Input of the package is the most essential for correct and accurate pick-up.

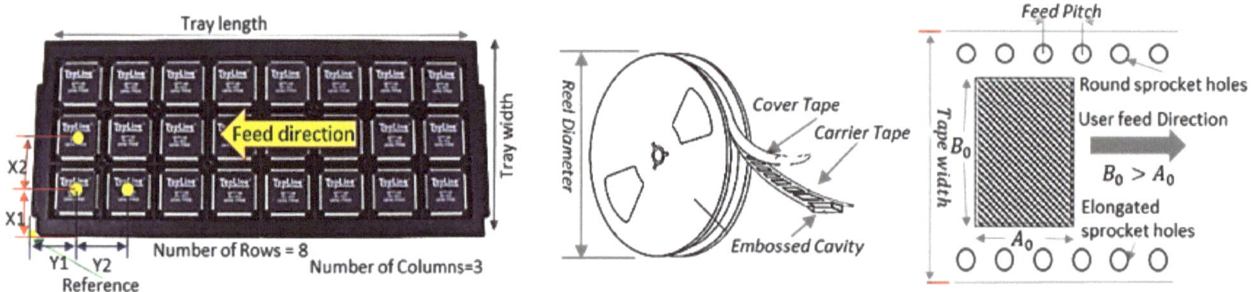

Template for creating Shape and Package Data

Data Input for Creating Package Data

$< Package\ type - Reel >$ $< Package\ type - Tray >$

- Tape width
- Feed pitch
- Real diameter
- Index speed
- Assign Feeder In
- Tape depth
- Tape check before
- Carrier Tape color
- Cavity (Offset X)
- Cavity Center (Offset Y)

- Pocket center location to Edge X
- Pocket center Location to Edge Y
- Pocket to Pocket center distance X
- Pocket to pocket center distance Y
- Number of rows
- Number of columns
- Total number of pockets

Data Input for Creating Shape Data

- Shape Data
  Import (Global, Job, Files) / Generate New-Shape Name
- Shape name
  Base information < Vision type, Switch image >
- Body
  Length, width, Height, Tolerences, color, lead brightness, 1st pin position X and Y, Bend position
- Lead
  Quantity check, pitch, telerences, length, width, thickness, check point
- Pin
  Vision Height Offset, Area offset, side light
- Error
  Alternate feeder, dump position, Force recover, Recover time
- Nozzle
  Min Dia, Max dia, Name. Enable wide Range
- Pick
  Do offset, Pick check mode, Offset X, offset Y, Offset θ, offset Z

Two options for creating the Part data for a component. Relational and non-relational. Normally the relational method is a default method.

Non-Relational mode

When this mode is used, the "Part Number", "Shape Name" and "Package Name" data is all saved under the one file name. There is no interlinking between the different parts of the data. This mode prevents users from accidentally editing packages and shape data for various parts at one time. With this mode active, only one component is used for creating and editing part data.

Relation Mode

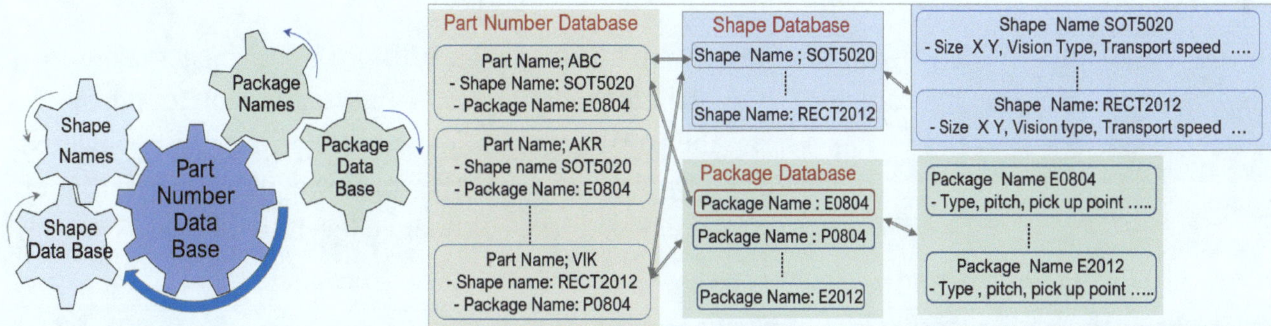

In the Relation mode method, the part number will be linked to a shared 'shape name 'and 'package name". This allows the multiple "part number" to use the same shape name and package name. Hence, by editing a packaging or shape record that is accessed by multiple part numbers, all the part data for all the part numbers are updated at one time.

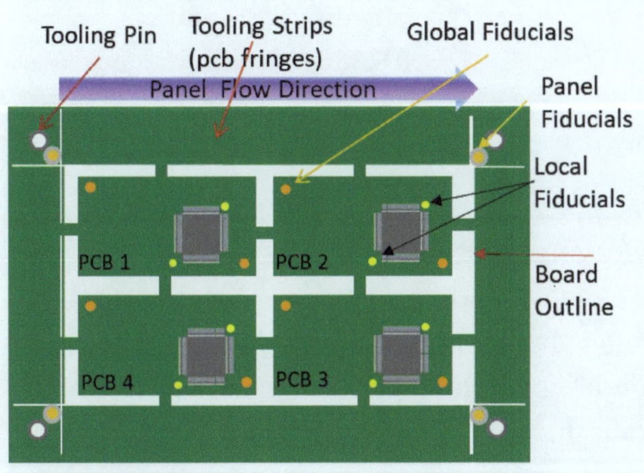

Mark Data

Also called Fiducial data. The programmer needs product PCB fiducial information about the type of fiducial pattern, diameter, color, center color, isolated or non-isolated, etc. so that the vision system can recognize the mark.

A Fiducial is a board feature used for global and local error correction to determine the difference between programmed coordinates and actual locations on the board. This ensures that parts are not placed on the respective pattern before their locations are verified.

Fiducial marks provide common measurable points for all steps in the assembly process. This allows all automated equipment to locate the circuit pattern. Fiducial marks are generally categorized into three categories. Panel fiducials, Global and Local fiducials.

Global fiducials are used to locate the position of all features on an individual printed circuit board. Local fiducial marks are used to locate the position of individual land patterns (or components) that may require more precise location such as 0.3 pitch QFPs, micro-BGA, etc.

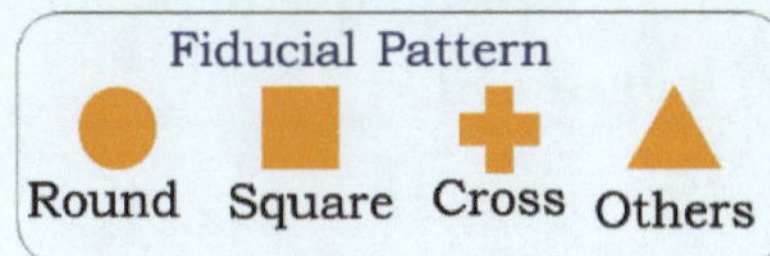

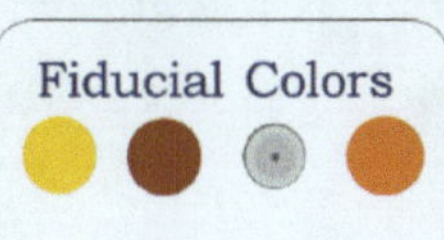

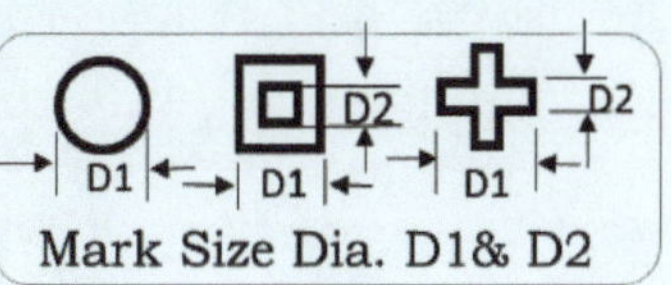

The most typical types of Fiducial failures are caused by improper color, size of Fiducial, and lighting values. Other factors such as the Confidence level and Search area can also be trouble spots, but as the programmer's experience level increases, these will be less likely to cause problems.

Mark Image

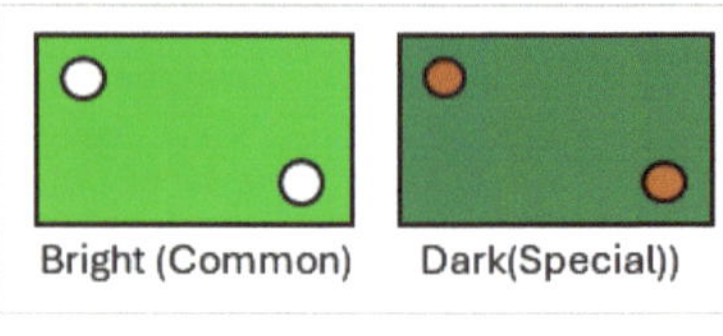

"BRIGHT" or "DARK" can be selected for a mark image taken by a camera. If fiducials or pad sites are consistently not found by the vision system, the parameter confidence level value can be lowered. If the vision system finds objects other than the fiducials or pad sites, the programmer increases the confidence level.

Another factor that the PCB Designer / Programmer needs to observe is the location of Fiducials. Location of Mark (The edges of the fiducial marks must be at least 4 mm from the board edges so that they do not get obscured by conveyor top clamps in the processing machines with fiducials-alignment camera systems. If the designer does not have the scope to put connectors, LEDs, etc., other than PCB Side edges, then the best alternative is to provide tooling strips along the sides of the Printed Circuit Board.

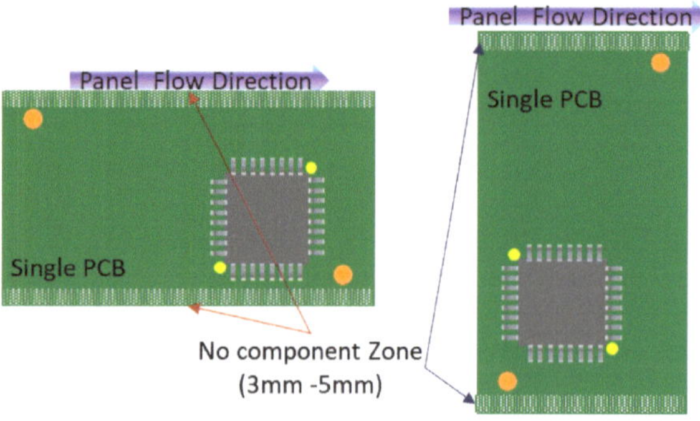

Search Area

The search area is one of the factors where the programmer needs to set a small window as much as possible to avoid wrong mark /fiducial recognition. When defining a search area, keep in mind that it should be large enough to allow some tolerance in board handling, but not so large that additional board features are found instead of the Fiducial or pad.

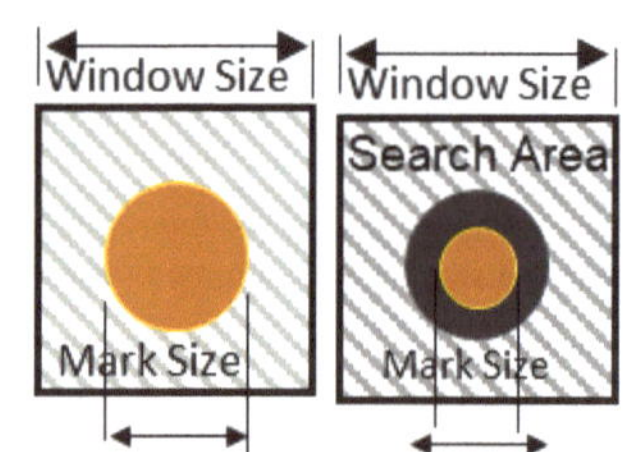

PANEL INFORMATION / PCB DATA

PCB Data is mandatory for the Placement of machine operation data and pattern programming. PCB length, PCB width, and PCB thickness are important parameters for the programming. Machine conveyor auto width adjustment happens according to the given PCB width measurement. The machine decides the PCB stop position at the work area based on the given board length measurement. Component placement height adjustment happens as per the given measured values of PCB thickness. A machine can check whether the programmed centroid data fits under the provided PCB data. For thin PCBs, the use of Pallets to support such PCBs is quite common in the manufacturing Industry. In that case, Pallet thickness needs to be mentioned.

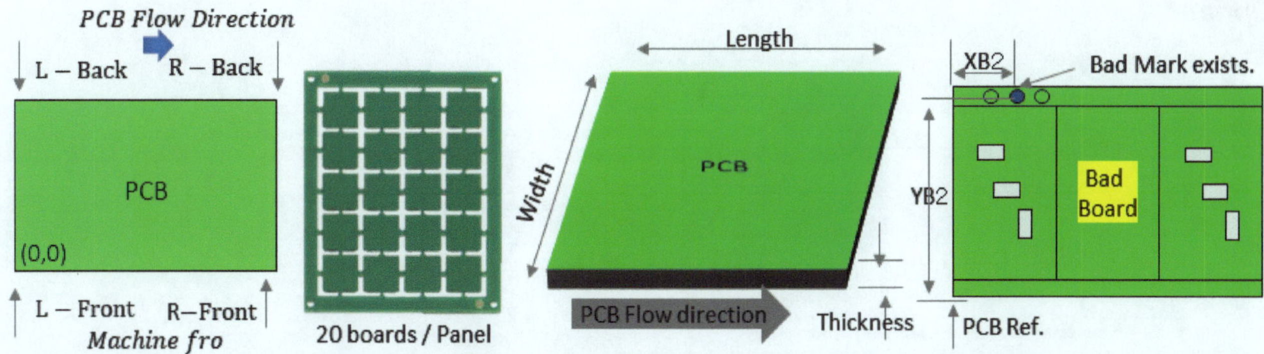

Board origin

The board's origin is defined as the (0, 0) value and is located at the lower left corner of the board from a top point of view. Even the reverse side of the board uses the lower left corner as the origin's reference point. Board origin may vary from the Pick and place manufacturer to the Manufacturer.

Panel Bad Board

In some cases, Bare board PCB Manufacturers produced a lot and later found a defect or series of defects in a PCB in the PCB array. The defect can be either a design or process defect. In such circumstances, the corresponding x-out fiducial will be marked on the PCB by the Manufacturer or by the user. Each **X-Out** board is to be marked in a manner so that the Pick & Place equipment fiducial camera recognizes bad marks and will not attempt to insert components on those boards. Programmers need to enable the BBR (Bare Board recognition) option of offline programming software.

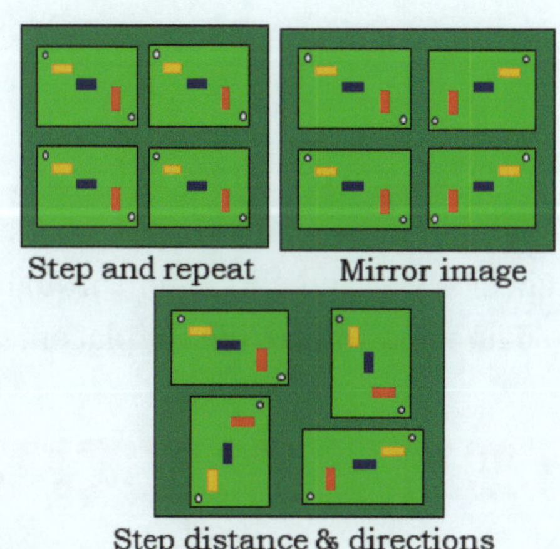
Step and repeat   Mirror image

Step distance & directions

Panel Information

Apart from panel length, width, and thickness, the placement program needs other data – Number of boards in a panel, Host interface, conveyor speed, and direction of boards within the Panel, such as Mirror image, Rotation, and flip boards. Programmers need to provide step distance offsets or step and repeat offsets.

PICK AND PLACE PROGRAM GUIDELINES

Before knowing the various terminologies of the data, one should know the basic flow diagram for creating the jobs for production. Here are the typical Fuji Flexa program guidelines followed by FUJI OEM for generating the recipes. The following Data is required and needs to be observed for any component placement programs. The programmer shall consider program optimization based on the Resource availability. I.e. type of feeder, quantity availability, type of nozzles, and available quantity.

Optional Data

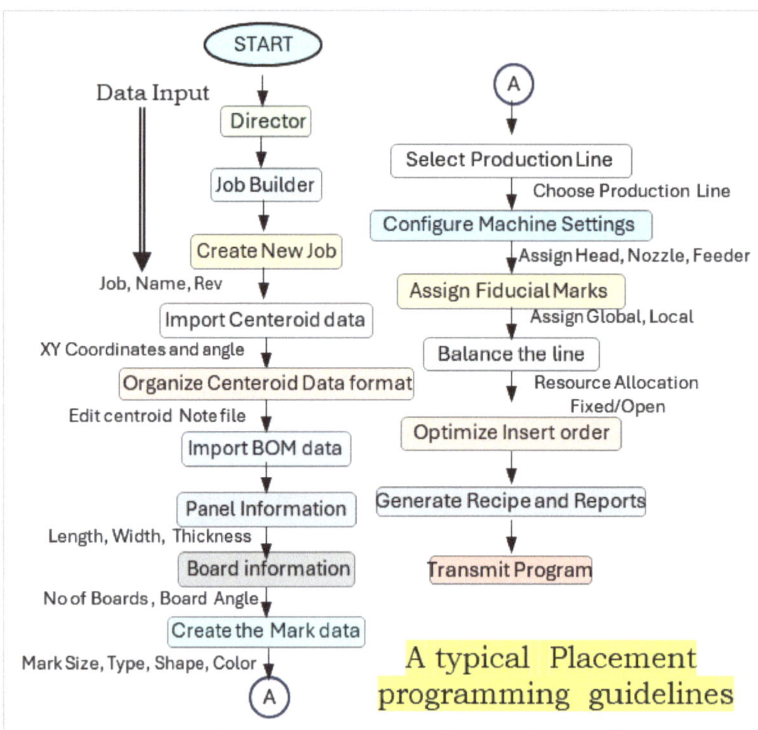

Since present technology pick and place Equipment is capable of numerous features and capabilities such as Coplanarity checks, glue dispensing, flux applications for POP (Package on Package), 2D checks, Cut and clinch for through hole insertion, die placements, etc., additional optional details /data need to provide for placement Programming.

PROGRAMMING CONCEPTS

Component placement Angle

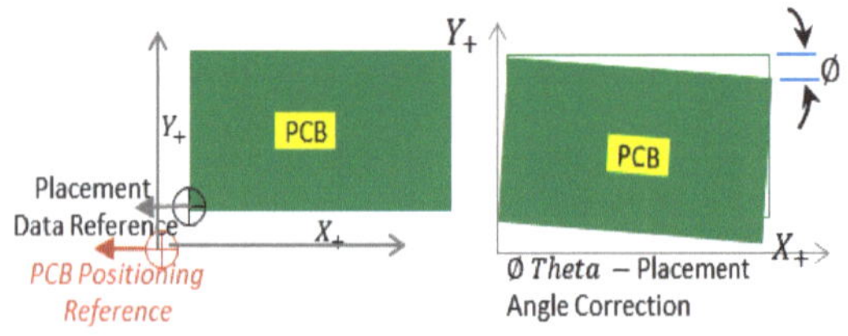

Automated pick and place machines are normally designed for machine reference, with the Board origin as the Lower left front. If cad data coordinates refer to different board origins, in such cases, the Origin offset must be provided in the program so that X, Y, and θ coordinates get aligned with the actual correct placement.

Case 1

Centroid data provides X-Y coordinates about the board origin and placement angle regarding standard packaging posture. The location is measured from the origin to the component center. In some rare cases, components can be supplied in a reel in different non-standard packaging postures. In such cases, the programmer should correct the placement angle in centroid data to flow direction.

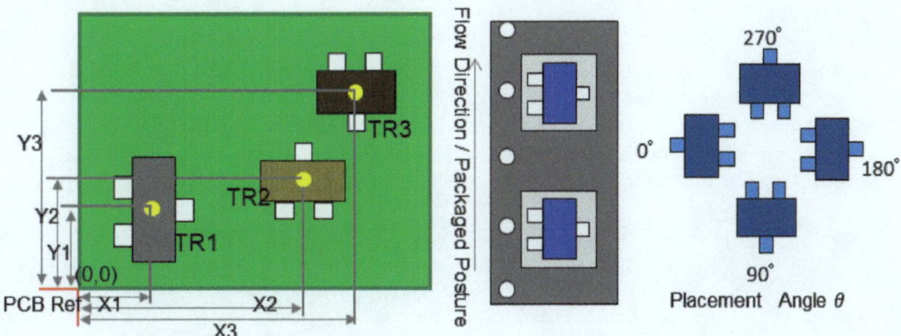

Mark Data

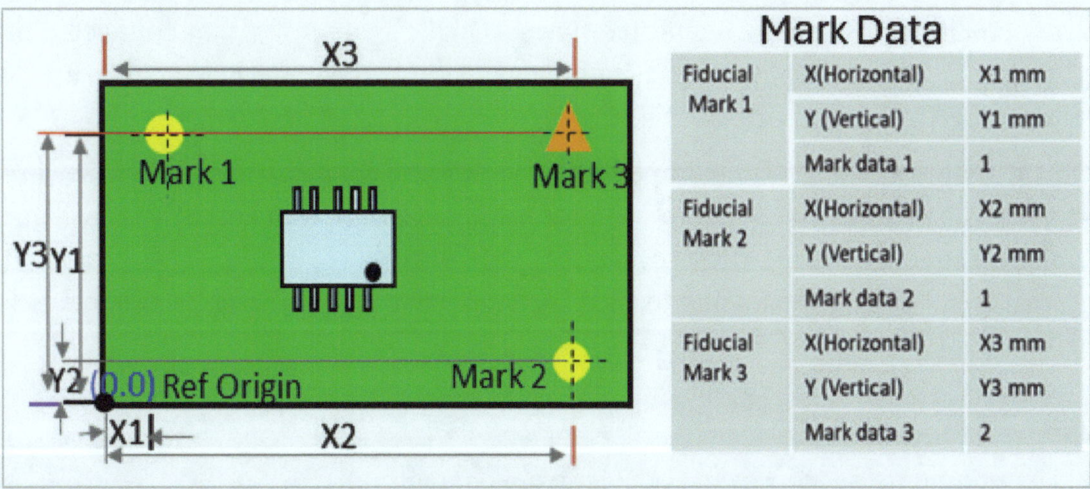

Mark Data		
Fiducial Mark 1	X (Horizontal)	X1 mm
	Y (Vertical)	Y1 mm
	Mark data 1	1
Fiducial Mark 2	X (Horizontal)	X2 mm
	Y (Vertical)	Y2 mm
	Mark data 2	1
Fiducial Mark 3	X (Horizontal)	X3 mm
	Y (Vertical)	Y3 mm
	Mark data 3	2

In this example, the PCB board origin is considered as Left Front. Always best practice to take two fiducials diagonally. The purpose is to correct the placement position XY and using Fiducials. 3Pt fiducials are more accurate than 2Pt fiducials, but cycle time matters a lot. In this example, the geometry of Mark 1 and Mark 3 are the same, and Mark 2 is different. The coordinates and mark data of the Fiducial mark are designated in the placement data.

Placement Pattern: Case I

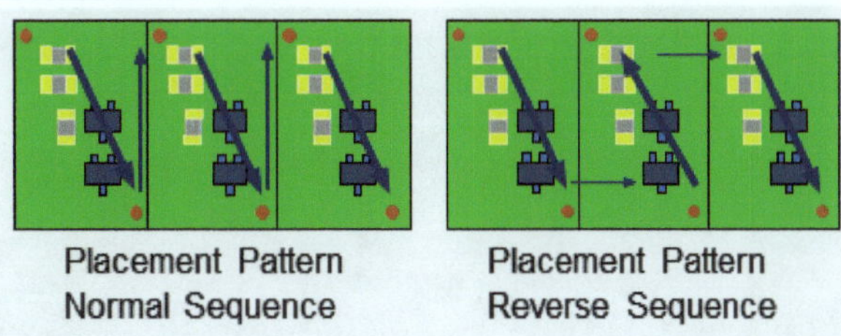

Placement Pattern
Normal Sequence

Placement Pattern
Reverse Sequence

Using the placement sequence normal and then reverse reduces the cycle time, and feeder/ Head movement is reduced.

Placement Pattern: Case II

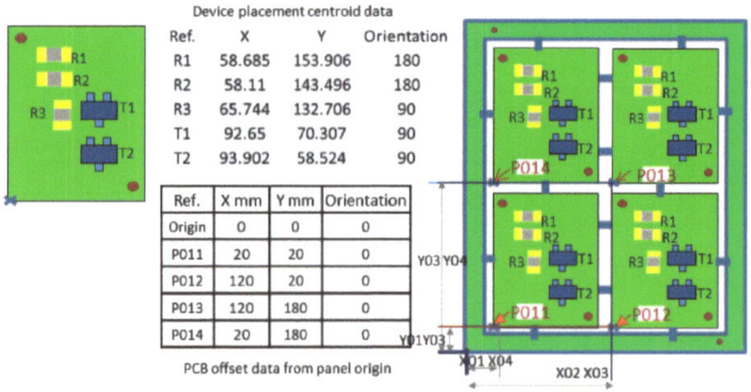

Device placement centroid data

Ref.	X	Y	Orientation
R1	58.685	153.906	180
R2	58.11	143.496	180
R3	65.744	132.706	90
T1	92.65	70.307	90
T2	93.902	58.524	90

Ref.	X mm	Y mm	Orientation
Origin	0	0	0
P011	20	20	0
P012	120	20	0
P013	120	180	0
P014	20	180	0

PCB offset data from panel origin

A PCB array combines a single PCB multiple times to make a larger array of connected boards, a "matrix," to speak. Repetitive placement data for each pattern. That means placing patterns P011, P012, P013, and P014 in sequence.

In this example, there are five components /boards and four boards /panels. So, a program can have twenty placement sequences in centroid data + four fiducial sequences. Instead of defining all 20 sequences, make the program only for one board, and the rest provide data for panel information, i.e., set information for P011 from panel origin ref, and for other panels P012, P013, P014.

Advantages

1. The number of sequences in Cad /centroid data minimized.

2. Any changes in one board are automatically corrected in all other boards. Corrective actions are quite easy.

Placement Pattern: Case III

A PCB Array combines a single PCB multiple times to make a larger array of connected boards, but in different Angles. Repetitive Placement component data for each pattern. i.e. Placing from pattern P011, P012, P013 and P014 in sequence are same.

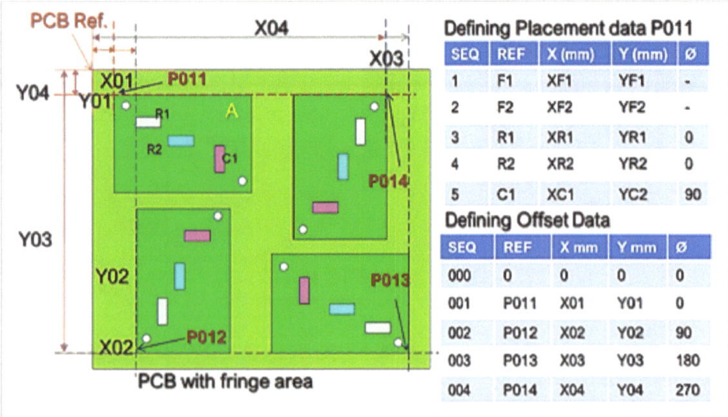

Defining Placement data P011

SEQ	REF	X (mm)	Y (mm)	Ø
1	F1	XF1	YF1	-
2	F2	XF2	YF2	-
3	R1	XR1	YR1	0
4	R2	XR2	YR2	0
5	C1	XC1	YC2	90

Defining Offset Data

SEQ	REF	X mm	Y mm	Ø
000	0	0	0	0
001	P011	X01	Y01	0
002	P012	X02	Y02	90
003	P013	X03	Y03	180
004	P014	X04	Y04	270

PCB with fringe area

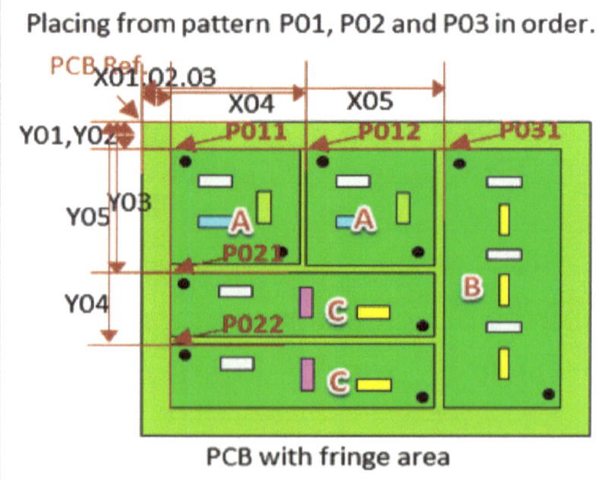

Placing from pattern P01, P02 and P03 in order.

PCB with fringe area

In this example, Placement coordinates are the same for each board concerning their board origin. However, board orientations are different in the panel. In such cases, define placement data for all components for board "A" having pattern reference P011. Define Input data for panel information – Offset information for P011 from origin reference and for other panels P012, P013, and P014. Angle reference. The same process is applicable for mirror imaged boards in a panel type.

Placement Pattern: Case IV

Multi-Model Repetitive Pattern Data.

In this case study, the Panel has three types of boards.

Product 1 – A (P011) (P012)

Product 2 – B (P031)

Product 3 – C (P021) (P022)

Programming instructions could be –

Input component placement data for product A.

Input Pattern reference data for P011 and P012. i.e., the distance between PCB ref to P011 and P012.

Input component placement data for Product B, Input offset coordinates from PCB ref to pattern P031.

Input component placement data for product C.

Input Pattern reference data for P021 and P022.

Offline software creates placement sequences and further process steps with the help of pattern reference origins.

SUMMARY

In this context, it describes the basic concept of placement programming. This is one step of placement programming, whereas optimization, cycle time, and line balancing are another step of programming. For cycle time analysis, the programmer should know available resources like feeders, nozzles, and machine capabilities. Cycle time analysis is the yardstick for production management, either to utilize or enhance the machine efficiency.

Reflow Process

ABSTRACT

The heart of the PCB assembly manufacturing process is only the "Soldering process". Soldering is a process that makes an interconnection between the component termination and the attached pads on the Printed Circuit Board. Soldering methods can be soldering automation, manual soldering, laser, and ultrasonic. Manual soldering can be done either with an iron tip or robotic soldering, which is widely being used in the present era of printed circuit assembly manufacturing. Reflow soldering, wave soldering, and selective soldering methods, or a combination of these methods, are the most popular and widely adopted Soldering automation methods in electronic manufacturing industries. The optimal soldering method must be selected by the bond strength, reliability, cost, and application requirement.

The process performance of the Printed Circuit Board Assembly (PCBA) production line significantly affected three major interconnected process segments for SMT, i.e., solder paste application, Component placement, and Solder Reflow. These three processes in combination significantly affect the ultimate solder quality. Other than the above three, process segments like inspection, testing, PCB separation, conformal coating, etc., enhance the product output quality, not the solder joint.

The reflow process means applying a temperature that is sufficiently consistent for the solder to flow evenly onto all the surfaces to reflow the solder paste to form a liquid solder volume, which is enough to create a perfect solder joint. Once the solder joints are formed, the assembly needs to be cooled to make it possible to handle the board for further processing.

A reflow machine is primarily used for soldering surface-mount components on a printed circuit board. In the PCB assembly production line, initially, bare PCBs are passed through a solder paste printer where, in the printing process, the solder paste is applied on pads through a stencil, Subsequently, paste-printed boards are transferred to the Pick and Place Machines where electronic components are mounted on the PCB. Now, these populated boards are sent to the Reflow machine to solder the components on PCBs by heating the board to a specific temperature profile that melts the solder paste and creates an intermetallic bond between components and the Circuit on the PCB.

The reflow process is one of the Critical processes in the PCB assembly production line. Populated PCBs are fed into the Reflow tunnel through a conveyor. Thermal profile is the most important factor for reflowing high-volume PCBs. Critical processes require proper time and temperature. In the reflow oven, these include the heat transfer rate of each heat zone and the conveyor speed.

A solder reflow requires strict temperature controls. Hence, a typical reflow oven is considered a good design when the oven delivers excellent thermal performance combined with Process Capability and Control. Hence, reflow oven OEMs must consider incorporating well-designed, calibrated Heating zones, cooling zones, and smooth-running conveyors. Of course, the number of zones, zone width, conveyor design, and nearly all other features provided by OEMs vary from model to model, depending on customer requirements.

A perfect thermal profile involves a spread over segments of temperatures on the PCB assembly with specified time intervals. It consists of heating the assembly boards and components held by solder paste through successively higher temperatures. A reflow profile needs to be created for the selected Solder paste characteristics, the density of components populated, and the type of PCB.

REFLOW OVEN – THERMAL PROFILE IMPORTANCE

Why Profiling?

- To remove flux volatiles

- To start flux activation to bring the materials being joined to a temperature that is sufficiently **high** for a final 'spike' of heat to perform the soldering operation

- Bring the materials being joined to a temperature that is sufficiently **consistent** for the solder to flow evenly onto all the surfaces to reflow the solder paste to form a liquid solder volume sufficient to create a sound joint

- Finally, the assembly is cooled to solidify the solder joints and allow board handling.

Hence, the performance of a reflow soldering oven depends on the design of the heating element that controls the system's heat, feedback or closed-loop system, the Conveyor section that controls vibrations, the PCB handling, the exhaust that controls heat in the chamber, and the cooling section.

MACHINE OVERVIEW

The selection of the Reflow machine depends on the company's objectives of a manufacturing unit and the quality acceptance criteria of a product that the customer is willing to implement for his manufacturing plant. Hence, the selection of Reflow Over is based on (1) Production volumes so that conveyor length is chosen accordingly. (2) Heating method – IR infra-red, Forced convection, or Vapor phase. Forced convection ovens are common and well-accepted in the present era of PCB assembly. Whereas IR and vapor phase ovens are being used for specific process requirements. (3) Process atmosphere – Reflowing PCB assembly in the presence of Air and Nitrogen is quite common, and the Vacuum method is optional. Reflowing the PCB in a Nitrogen atmosphere is mandatory for certain Class III products such as aerospace, medical, etc., and the presence of Fine Pitch components in the PCB assembly.

Reflow ovens use programmable heaters to generate heat. A program with all zone temperatures, as well as belt speed, is referred to as "The Recipe". The temperature that the board "sees" as it passes through the oven is called the "profile". The recipe is optimized to achieve the profile. Heaters are fitted in the reflow oven on the top side and bottom side, and the PCB passes through the conveyor in between throughout the conveyor length. The thermal profile for good soldering requires four profile zones, i.e., preheat zone, soak zone, Reflow zone, and cooling zone. The profile programmer allocates a certain number of heaters for each zone so that PCB passes between these top and bottom heaters to a time interval at set temperatures of each heater to fulfill profile requirements. Other process control variables – selection of solder paste, i.e., solder past characteristics, type, size, and PCB surface finish- also need to be considered for the perfect solder joint. The advantage of more heaters and coolers in a reflow oven is that programmers get the flexibility to align the process width and scope of production volumes for lengthier conveyors.

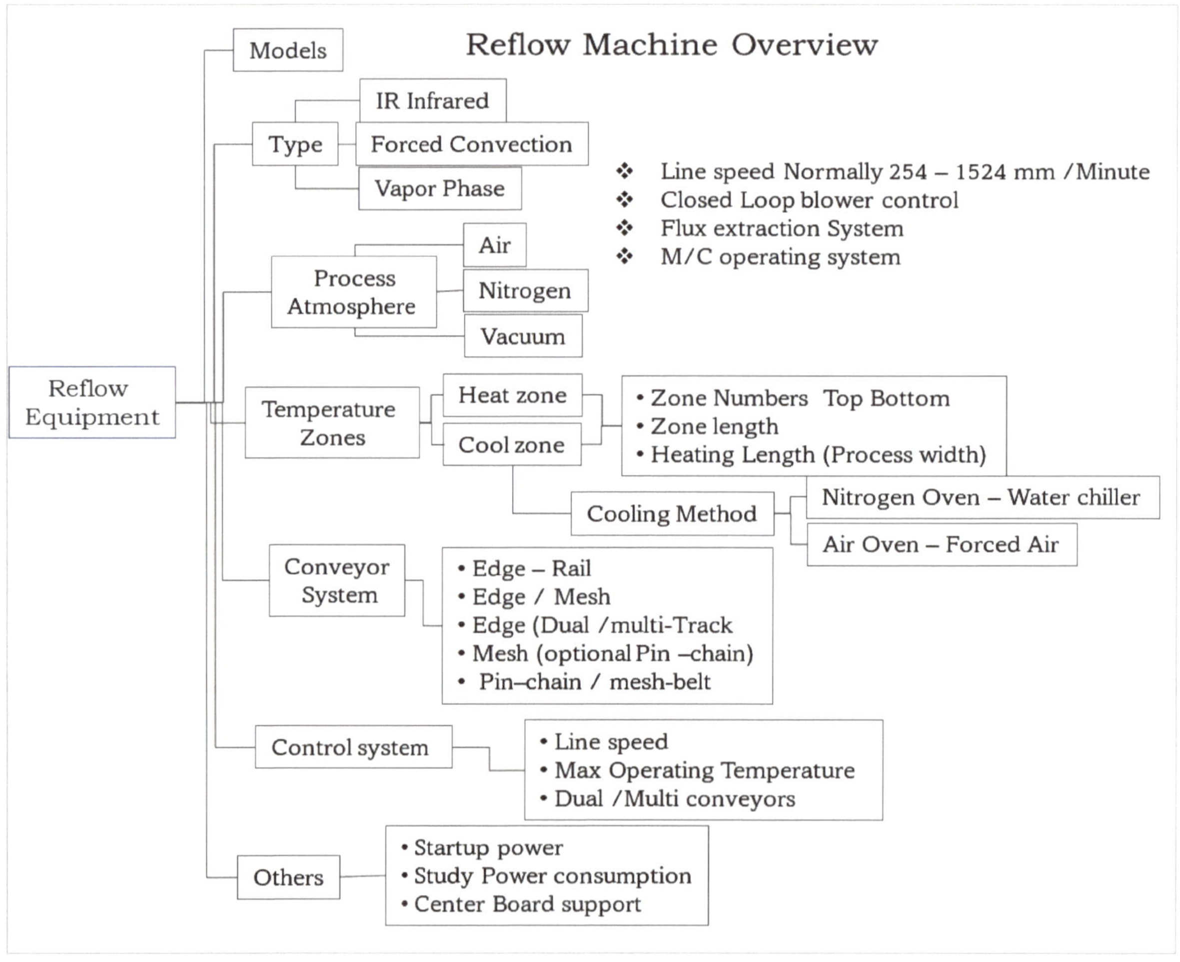

Reflow oven models are available in 5 heater Zones, 7 heater zones, 9 -10 Zones, and even 14 zones of reflow. Improvements and performance of machine components are one part, and zone configurations can be another part. Users are always stuck with the question of which reflow oven best fits their needs. To get the answers, one needs to consider the factors of Solder paste rheology, heated length of reflow oven, and production volume. Also, reflow ovens should be capable of process considerations for lead and lead-free environment, and challenges like Solder joint under Nitrogen atmosphere instead of Air.

Reflow ovens come with single conveyors, double conveyors, and even three conveyors. Single and double conveyor systems are common in the electronics industry. Making profile and heater temperature settings for zones is easy for a single conveyor system as the only type of PCB reflow happens, and the PCB temperature absorption characteristics will be the same. Dual conveyor reflow ovens are used to reflow Top and bottom PCBs simultaneously using single profile programming—for example, cell phone products. Options are available to choose the conveyor type, such as mesh, pin, belt, or combinations.

Machine control systems are user-friendly and provide access to individual temperature settings of the heaters' top and bottom, coolers, and conveyor speed. The power consumption of a Reflow machine plays a key role in designing for OEMs. The lower Startup power and study power consumption while meeting all other performance requirements of a machine draw the full attention of machine buyers. Other than this, features like a closed loop system and options for reflow under a Nitrogen atmosphere, CBS center board support, and flux extraction system will always be an added advantage for the enhancement of overall performance.

UNDERSTANDING HEAT TRANSFER METHODS

Heat transfer is the exchange of thermal energy between physical systems by dissipating heat, depending on the temperature and pressure. The fundamental modes of heat transfer are conduction or diffusion, convection, and radiation. Temperature and the flow of heat are the basic principles of heat transfer. The amount of thermal energy available is determined by the temperature, and the heat flow represents the movement of thermal energy.

On a microscopic scale, a molecule's kinetic energy is directly related to thermal energy. As temperature rises, molecules increase in thermal agitation, manifested in linear motion and vibration. Regions that contain higher kinetic energy transfer energy to regions with lower kinetic energy. Heat transfer can be grouped into three broad categories: conduction, convection, and radiation.

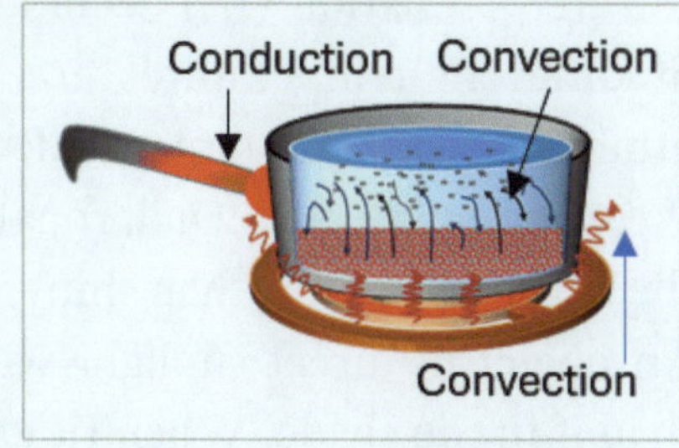

Conduction

Conduction transfers heat via direct molecular collision. An area with greater kinetic energy will transfer thermal energy to an area with lower kinetic energy. Higher-speed particles will collide with slower-speed particles. The slower-speed particles will increase in kinetic energy as a result. The process of heat conduction depends on the factors of temperature gradient, cross-section of the material, length of the travel path, and physical material properties. The temperature gradient is the physical quantity that describes the direction and rate of heat travel. Temperature flow will always occur from hottest to coldest or, as stated before, higher to lower kinetic energy. Once there's thermal equilibrium between the two temperature differences, the thermal transfer stops.

Convection

When a fluid, such as air or a liquid, is heated and then travels away from the source, it carries the thermal energy along. This type of heat transfer is called convection. The fluid above a hot surface expands, becomes less dense, and rises. At the molecular level, the molecules expand upon the introduction of thermal energy. As the temperature of the given fluid mass increases, the volume of the fluid must increase by the same factor. This effect on the fluid causes displacement. As the immediate hot air rises, it pushes denser, colder air down. This series of events represents how convection currents are formed.

Radiation

Thermal radiation is generated from the emission of electromagnetic waves. These waves carry the energy away from the emitting object. All materials radiate thermal energy based on their temperature. The hotter an object is, the more it will radiate. The sun is a clear example of heat radiation that transfers heat across the solar system. At normal room temperatures, objects radiate as infrared waves. The temperature of the object affects the wavelength and frequency of the radiated waves. As temperature increases, the wavelengths within the spectra of the emitted radiation decrease and emit shorter wavelengths with higher-frequency radiation

IR INFRARED HEATING METHOD

Infrared Radiation (IR) occurs when two bodies of different temperatures are in sight of each other. The best example of IR is the heating of the earth by the sun. Dull, rough surfaces absorb the sun's rays better than shiny, smooth surfaces. An object in direct sunlight will become hotter than if in the shade. When IR energy encounters a target, some of it is absorbed, some reflected, and some transmitted through the target. The absorbed energy heats the targeted material, the reflected energy is returned to the source, and the transmitted energy is lost.

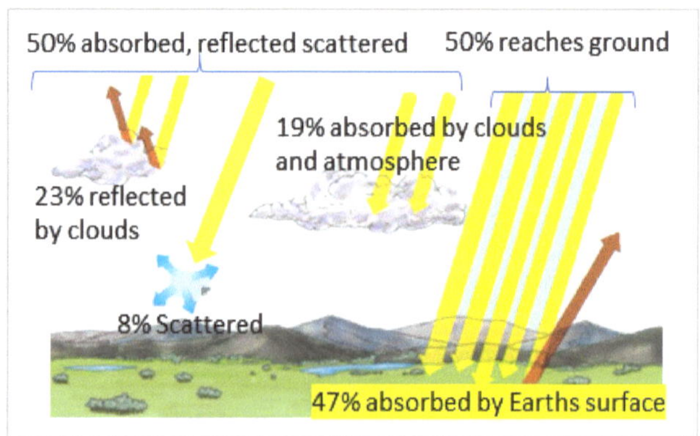

IR in SMT applications works similarly. Fluxes, plastic components, and epoxy glass laminate absorb IR very well. Shiny, reflowed solder will reflect the IR energy away. Solder joints around small packages (such as chip resistors, capacitors, and SOICs) are very well in sight of the IR energy and heat. Solder joints around larger devices (such as PLCCs, and QFPs) are shaded and do not heat as well.

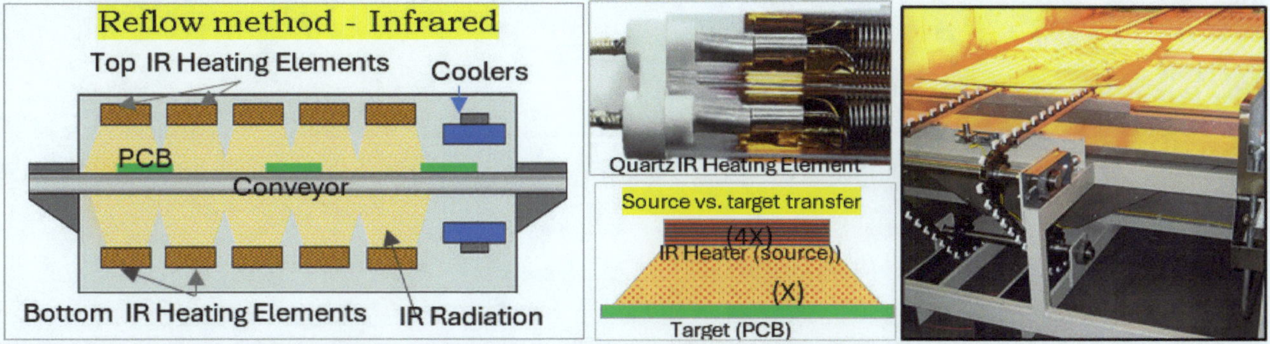

Infrared Reflow, Lamp Emitters Type

The heat source is generally from ceramic infrared heaters, which transfer heat to the assembly using radiation. Lamps (tungsten or quartz) are used as an IR energy source. Quartz glass is a preferred material for IR because it is transparent to IR radiation, can withstand constant working temperatures of more than 1000°C, and is resistant to chemical corrosion. Quartz with single or double tubes provides the best mechanical resistance.

Quartz tubes coated with Gold/White /ruby for industrial use as they reflect the infrared radiation and direct it towards the product to be heated. Consequently, the infrared radiation impinging on the product is virtually doubled. Gold is used because of its oxidation resistance and very high IR reflectivity of approximately 95%.

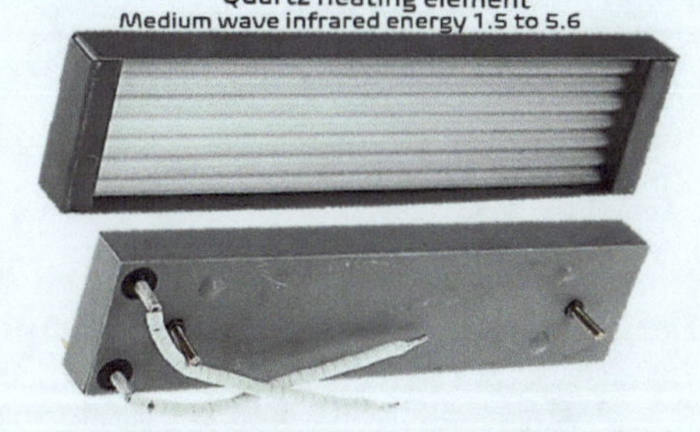

As the PCB travels to higher preheat zones and eventually to the soldering zone, the wattage is increased on the lamps, and the wavelength decreases. This increases heat transfer through radiation and decreases the convection component. The energy transferred by conduction through the PCB depends on the speed of PCB transfer. Much of the heat transfer is done from within the PCB, solder paste, or component. The majority of heat transfer is through radiation in the near IR wavelengths.

Advantages

Low mass, partly electromagnetic spectrum, Power, and wavelength are functions of temperature, hence fast response. Profiles can be changed to changes in applied wattage; as convection plays less of a part in heat transfer, the response time is much less. Instant ON and OFF leads to energy savings.

Disadvantages

Uneven heating. Creating profiles is very difficult as the system's response to wattage change is fast. Profile adjustments are possible only by changing the conveyor speed. Hence, product changeover time is slow. Profile stability is another concern because of the lamp emitters' aging.

The temperature differential between the (source)heater (4x) and the (target) board (x) is not directly related. If PCB requires 1(X) unit of heat absorption rate, then the IR heater must be set 4 times(X).

IR allows you to target exactly where you want the solder reflow to occur. Though this is a good point, the shadow effect is a drawback. Further, color-sensitive, Darker colors will absorb heat more.

[(Short wavelength infrared ranges from 0.76 - 2 microns or 7000°F - 2150°F.
Medium wavelength infrared ranges from 2-4 microns or 2150°F - 845°F.
Long wavelength ranges from 4 - 1000 microns or 845°F - below freezing.)]

CONVECTION REFLOW

Convection and forced convection technology dominated IR reflows because IR technology is not able to be fulfilled with complex boards because temperature control was always difficult to achieve due to variations in board layout, thermal mass, and reflection of the board and components. Long-term change is also one concern. The response of the system to wattage change is fast, and the profile adjustment can be done by changes in conveyor speed alone.

Forced convection reflow uses blowers on natural convection heaters and was developed during the late 1980s as an alternative to infrared heating. Forced convection reflow ovens, in which an external force pushes or pulls the flow over the object, are popular nowadays. Typically, forced convection heating or cooling rates are higher than natural convection rates. In this technique, the source emissivity and target absorptivity will be a 1:1 ratio.

When reflowing in a nitrogen atmosphere, the convection method became popular because blowing hot gas at the circuit provided a more controllable and consistent heating regime. Transfer heat through gases is limited by their low specific heat. So, convection systems always need a high gas flow. It would not be cost-effective to pump the gas only once, and practical systems always recycle gas in a closed loop within individual zones. Depending on the design, around 10–25% of the gas used will be sourced externally, with the rest recirculated.

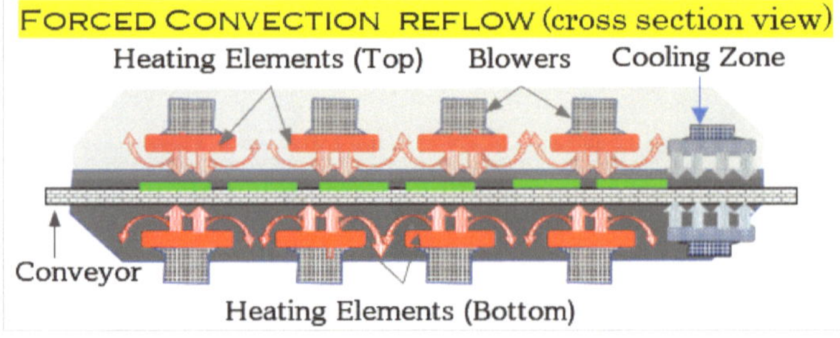

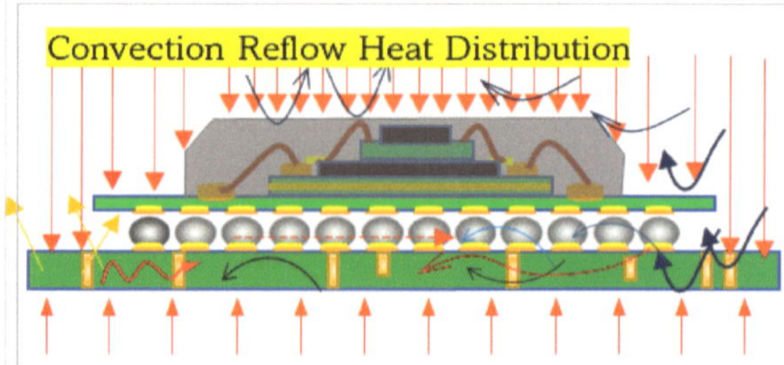

Equipment manufacturers integrate Fans, Compressors, or Flow amplifiers to generate forced convection in Reflow ovens as a standard practice. Some OEMs establish patented designs for better closed-loop power-saving methods. Most convection reflow ovens use fans for forced convection. Incorporating fans is an inexpensive, reliable method of moving a high volume of air. Exit velocities may be limited due to low-pressure generating capability.

Compressed input gas sources such as compressed air or Nitrogen are also used to create forced convection. Compressed gas reflow is used when the application requires an extremely accurate thermal profile with filtered process gas. It means that pressure must be regulated to achieve constant flow. Humidity can be controlled using an inline dryer.

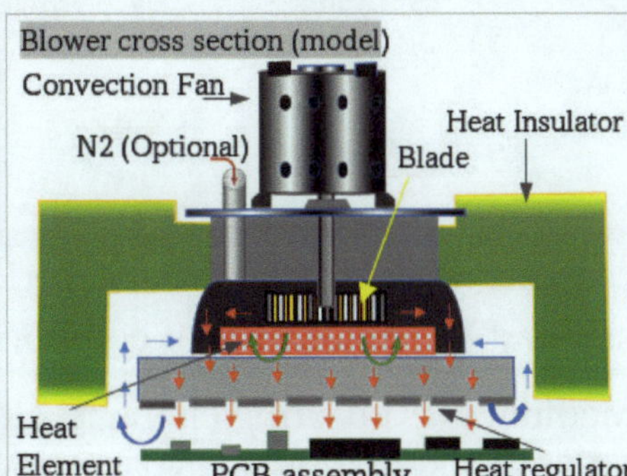

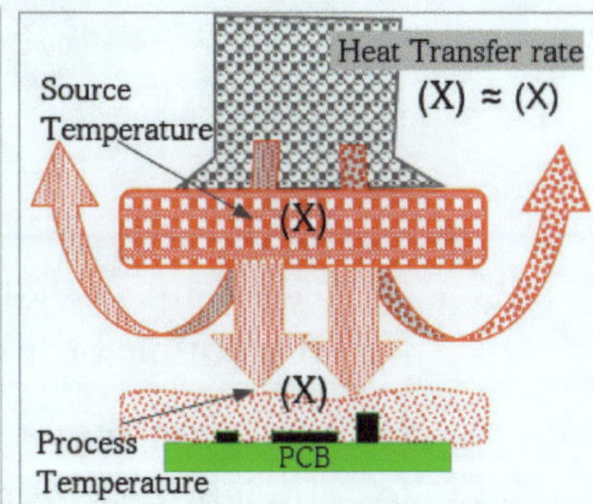

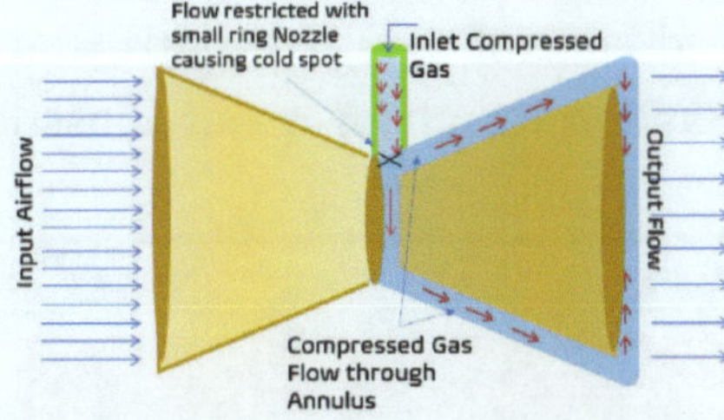

Flow amplifiers, on the other hand, use compressed sources to induce flow. Compressed gas flows through the inlet into an annular chamber. The gas is then throttled through a restriction at high velocity. The compressed gas flows toward the outlet, adhering to the outside wall. A low-pressure area is created in the center, which induces flow.

REFLOW HEATERS

Heater panel construction, design, and quality of the heating element are all factors that contribute to the performance of reflow ovens. Heaters need to be robust and designed to give uniform heat and maximum surface area to maximize heat transfer. It should respond quickly to any changes in profile requirements. Small turn on time to be maintained for reaching the set value temperatures.

Engineering innovations towards the Heating module or element type, fan design, and nozzle matrix for air flow brought the oven manufacturers to build various models by providing ovens with energy-saving management. Here is an example of Heat module design patented technology adopted by OEM (courtesy ITW EAE) oven manufacturers to improve uniform gas management, flux management, i.e., flux extraction system, and exhaust management.

- Isothermal Chamber Technology delivers internal compression zones that equally distribute the air across the entire diffuser. As a result, efficient heat transfer, excellent zone-to-zone segregation, Controlled environment for lower power and N2 consumption, and excellent separation between heating and cooling for better Time above liquidous state (TAL).

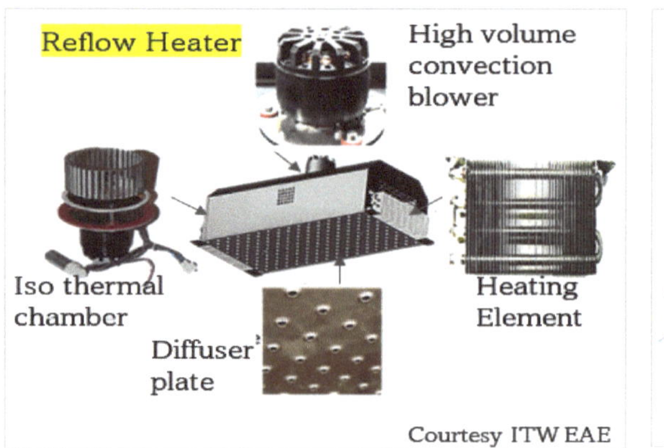

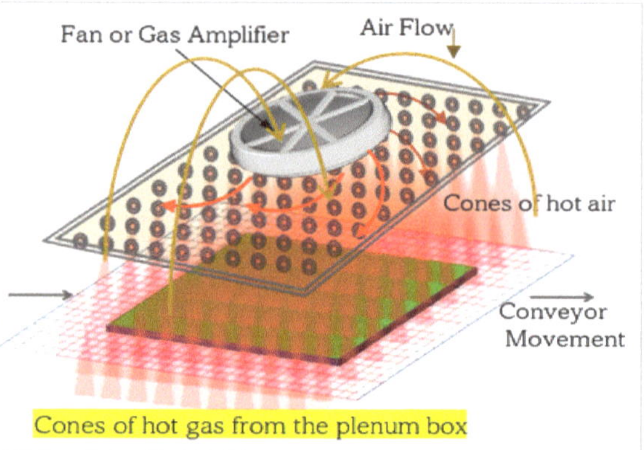

- Diffuser Plate - Balanced airflow, overlapping concentric cones of air, Envelop product for efficient heat transfer and NO movement of components

- Heater Technology - Medium mass heating element with steel fins. Large surface area for efficient heat transfer. Combined 900 sq in of surface area. Heat on intake to mix the process gas.

High volume convection Blowers - Independent blower speed control, Low, Medium, and High. Independent access capability, Individual plugs for power and quick change, and Closed loop blower control.

VAPOR PHASE REFLOW /CONDENSED SOLDERING

It is also known as condensed soldering. The vapor phase is generated by heating a dense medium fluid (normally PEPF perfluoropolyether) to a stable boiling temperature. The latent heat of liquid vaporization provides heat for soldering. The liquid chosen for soldering determines the peak temperature of the board.

Heat transfer is dependent on a temperature difference between the vapor and the device and stops immediately when both reach the boiling point of the medium. This process is suitable for soldering odd-shaped parts, flexible circuit boards, and tin lead and lead-free surface mount packaging leads.

This is a very good Technique to avoid heat shocks. Since latent heat of vaporization is used, temperature can be controlled precisely. Based on the heat capacity of the component, sufficient heat is absorbed by the component. The component temperature never exceeds the vapor temperature. Hence, no issue of Overheating.

The organic solvent PFPE (Perfluoropolyether) will be used in a container, which boils and generates vapors at 230°C. Hence, you can fix the entire process at 230°C. Ideal for lead-free. Ideal For BGA reflow soldering to avoid the Popcorn effect. Conducting energy uniformly without dependence on circuit assembly mass, geometry, color, or composition. Entrapment of moisture causes separation of solder balls if any from the substrate.

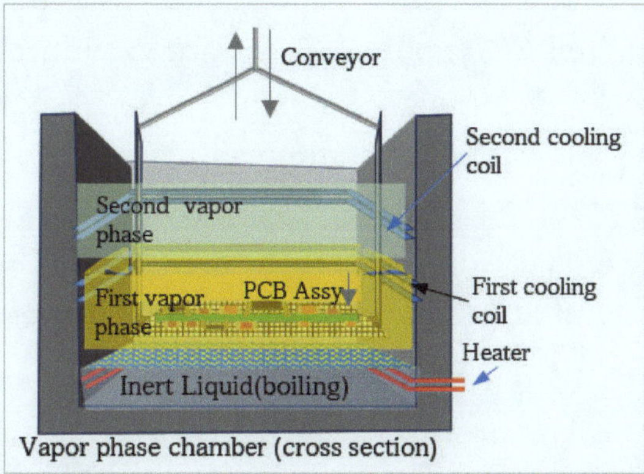

Vapor phase chamber (cross section)

Figure shows the structure of the equipment used in this method. This equipment consists of the first vapor phase used for the batch reflow soldering, a preheater, cooling, and a second vapor phase to prevent splashing of the liquid from the first vapor phase.

Process steps for Lead-free Soldering

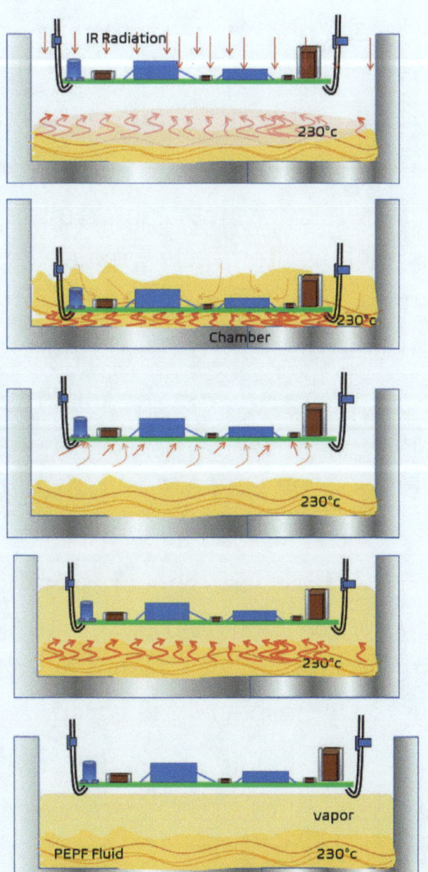

1. The assembly is moved to the vapor phase. It was preheated by IR radiators from the top. It also can be preheated on the upper boundary of the vapor

2. phase predominantly from below. The use of IR preheat improves the quality of the solder joint and allows the free shaping of the temperature profiles.

3. The assembly dives into the vapor phase and is heated up. A patented procedure allows for controlling the heat transfer to the assembly so that a soft temperature rise is performed.

4. The vapor condenses on the assembly and transfers its heat. Since the vapor is inert, it performs in an inert gas atmosphere with 0 ppm oxygen. This is done automatically without the use of nitrogen.

5. The assembly can be heated to a maximum temperature equal to the temperature of the vapor. This temperature cannot be exceeded even if the assembly stays longer inside the vapor. Therefore, overheating is not possible.

6. After leaving the vapor phase, there is still condensed fluid left on the board. Due to the inner heat of the assembly, the liquid evaporates, and a dry assembly leaves the machine.

Vacuum Reflow Process

It is a combination of conventional reflow soldering with a simple Vacuum process. Use of vacuum mainly for removing volatile substances such as voids from the solder joints.

Simple process – Solder of component melted→Component Goes into Vacuum chamber → Application of Vacuum → Voids sucked out of solder joint → Component Goes into cooling chamber. Adjustable Process parameters can be 1. Evacuation and 2, Holding and venting time. The pressure in the oven will be reduced to about 200mbar. The vacuum shall be pulled ≈ 30sec after the solder becomes molten

For profiling, the typical pressure curve (Pressure level and dwell time) and the specific temperature curve (precisely adjustable Heat and cool-down Ramps) can be controlled.

SOLDERING IN A NITROGEN ATMOSPHERE

Why Nitrogen?

Oxygen often reacts with the metals. If you leave iron in the atmosphere, it rusts because of oxygen. After the reaction, it forms metallic oxides. These oxides form much more rapidly in a hot atmosphere. This is the property of oxygen.

In the electronics industry, soldering is a major part of the process. Soldering involves metals. The wires are metal. Soldering iron tips are metal. The pads are metal. Even the solder paste itself is metal. The process requires temperature. I.e., all factors are simply against the property of oxygen. Hence, oxides form, and these oxides shield the conductor and try to keep the solder from adhering to what metal is left.

We use flux in the solder process. Flux eats the oxides, but it also leaves residues, which in turn react with Oxygen. The concept of miniaturization of components, development of high-density components, and the introduction of BGAs, CSP, POP, Fine pitch QFP, etc. raised concern about solder joint quality.

With these concerns and limitations, do we really want a lot of Oxygen around when soldering? Obliviously Not. Hence, flooding the inert gas is one of the solutions. Nitrogen is one of the inert gases, having a property of very slow reaction with metals, and is very much available in the atmosphere and cheap.

It is widely accepted in the electronic assembly industry that nitrogen 'improves' the wave soldering process and reflow process. However, the use of reflow equipment with Nitrogen is much less frequent.

Effects of Nitrogen

Based on experiments, research, and industry data, it is confirmed that the build quality and reliability of solder joints improved using Nitrogen. Drastic improvement found in defective joint DPM. Improvement in first-pass yield was witnessed. Other factors also strengthen the usage of Nitrogen.

Spreading behavior - Spreading starts at lower temperatures as we reduce the Residual Oxygen Level (ROL). Leaded solder is found to be spreading at 401°F (205°C) if the ROL is <10 ppm, but it needs 404.6°F (207°C) at 100 ppm and 518 °F (270°C) at 1000 ppm. The same pattern holds for other solders, indicating an inhibiting property of oxides to the tendency of spreading.

Good copper coverage by liquid solder is achieved only at low ROL (<40 ppm). No Dewetting was observed for the same experiment performed under nitrogen. Even very low levels (20 percent) of NV matter in the flux did not result in any dewetting. Comparing the wetting force for different fluxes in air and nitrogen again yields results indicating that nitrogen coverage can improve the process.

Wetting angle - Small wetting angles indicate good wetting behavior and usually indicate a sound joint. When measuring wetting angles for a variety of fluxes, researchers found that wetting angles were much smaller under nitrogen than in air for most fluxes. Thus, nitrogen improves wetting for most flux vehicles employed.

The surface tension of the liquid solder is important in the soldering process. Measurements again indicate that the surface tension is higher under nitrogen than under air. Surface tension will be lower under air than under nitrogen.

Properties Of Nitrogen

1. Nitrogen makes up about 78 percent of the air; 20% is oxygen; whilst the rest consists of CO; CO2; and a few of the 'noble' gases, primarily argon; Nitrogen in and of itself is not toxic, but of course it cannot sustain earthly life, which is based upon a carbon-oxygen cycle.

2. Nitrogen boils at about -320.4°F (-195.8°C) (normal pressure), so liquid nitrogen as it may be delivered is very cold.

3. Cold nitrogen also is heavier than air and thus may accumulate in low-lying areas.

4. Nitrogen is not manufactured but rather is extracted from the air. The nitrogen extracted from the air is returned to it.

5. There are several different ways of extracting nitrogen from the air. The most common method is a 'freezing' process, usually referred to as 'cryogenic.' A cooling process is used to liquefy the gas, which in a kind of reversed distillation process purifies itself because of the different boiling points of oxygen and nitrogen.

With these different processes, it is possible to provide nitrogen from a purity of 95.0 to 99.99999 percent [=0.1 parts per million (ppm)], depending on the need of the process. Since most assemblers require a high-purity atmosphere (<500 ppm), only liquid nitrogen (cryogenic purity) or Liquid Assist onsite systems (cryogenic purity) are recommended for assembly.

BOARD TRANSPORT SYSTEM

The board transport system may use either a conveyor belt or an edge conveyor system, often with a mesh belt underneath. A chain conveyor has pins protruding from the sides of the chain (Approx 4mm) to provide edge support that carries the board through the reflow system without touching the belt. This allows double-sided PCBs to be processed without disturbing the bottom-side components.

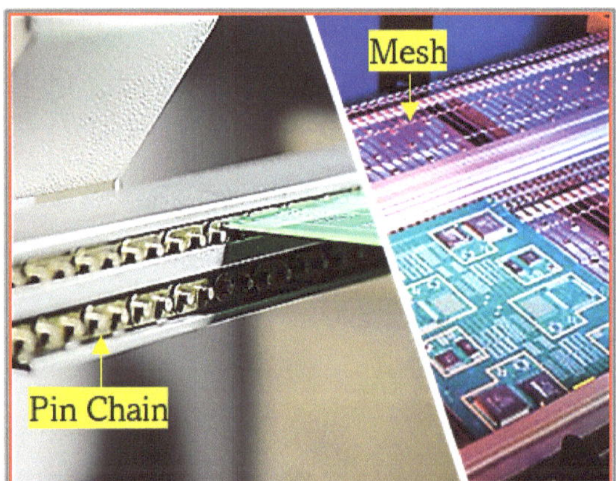

A reflow conveyor rail assembly is made with a heavy-duty cross shaft, often a single mold capable of compensating for thermal expansion. The proper design of the edge conveyor system is critical, as typically only 4 mm is in contact with the bottom-side edges of the product. Edge conveyor straightness is a key issue in SMT processing. Else boards fall. In fact, some suppliers even go to the extent of building in a tensioning system to pull the rail assembly straight and counteract the tendency to grow and warp during the machine's warm-up period. In-built oilers are common as a part of conveyor design. Oil drops dip on moving conveyor at rated frequency to lubricate pins at conveyor input.

CENTRE SUPPORT SYSTEM

This supports the center of the board to prevent sagging during reflow, which means either using thermally inefficient board carriers or devising a more complex conveyor system. Panels that are larger and heavier, yet thinner, require a special PCB design where there is a small clear path on the center of the board to avoid contact with tensioned moving wire in CBS.

COOLING ZONE

The cooling system aims to ensure that the product is below the liquidus temperature before exiting the oven. An initial cooling jet between the reflow zone and cooling stage can dramatically reduce the time above liquidus, providing a tighter grain structure and better solder joint.

The main product cooling, however, is usually based on a water-to-atmosphere heat exchanger, where gas convection cools the product from both top and bottom. Issues relating to this stage include a) dealing with the build-up of flux residues, b) keeping the exit temperature of the product sufficiently low, and c) reducing the gas consumption (where nitrogen is used).

FACTORS TO BE CONSIDERED WHEN CHOOSING THE REFLOW

- Reflow ovens with 4 -6 heating zone models are best suitable for curing applications. These are not suitable for high production volumes, and are not capable of high-density complex boards.

- Reflow ovens with 8 to 10 heating zones and long ovens with faster belt speeds can be the best solution in high-volume manufacturing environments. Wider reflow profile process window achievable for reflow ovens with more than 10 zones.

- Solder paste manufacturers generally provide a fairly wide window (in terms of total heating time) for the various stages of the reflow profile: 120 to 240 seconds for preheat and soak and 60 to 120 seconds for reflow/time above liquidus. The Solder Paste Heating Time factor needs to be considered.

- The next consideration is the total heated length of the reflow oven. It is very important to differentiate between the oven's total overall length and its heated length. Almost all oven manufacturers will provide the oven's heated length, sometimes referred to as heating tunnel length, in their specifications.

- Cooling is important but typically doesn't affect soldering quality unless the PCB is cooled too quickly.

- CBS, Center board support system, multi or triple conveyor, and Nitrogen atmosphere are all oven options, and these features are provided by the Oven supplier on demand. The oven buyer should have a vision about his business and the scope of assembly product while choosing an additional investment for options.

SUMMARY

DFM (Design for manufacturability) is the most important factor for the success of a product. An electronic product designer always designs his product in terms of PCB design, selection of Components, packages for his circuit, and keeping the given Process design. Product design versus process design, means that during /finalizing product design, process specifications also have been designed and decides on what type of Solder Paste need to be used, PCB substrate selection, PCB surface finish, the requirement of Center support system so that the product should meet standards of Reflow Oven for manufacturability. If the Product PCB is loaded with Heavy components, the PCB is designed in such a way that a clear path is provided for the Center support system in reflow. The key to the success of soldering is the 3:1 rule. When considering soldering standards issued by IPC-JEDEC, the ideal profile follows the 3:1 rule, i.e. 3 parts pre-heating to 1 part reflow.

The number of zones within the reflow process does not influence the ΔT. Air volume is the governing factor, and by using very powerful and highly controllable fans, SMT can achieve the lowest ΔT across any PCB. New versions of Reflow oven models are ROHS complaint, where heating element modules are designed to sustain reflow peak temperatures of $255\pm5°C$. The success of reflow Oven technology is purely dependent on Thermodynamics applied in product design and operating cost. Users of a reflow oven are more concerned about the power consumption and running cost of Nitrogen.

Reflow profiling

ABSTRACT

For a layman's understanding, consider an example of a homemaker making Rotis using a Tawa and a Gas stove. If you observe the process: mixing ingredients with water—keeping dough wet for some time—using a rolling surface and rolling pin to flatten dough balls by hand—beginning to cook the roti on a Tawa for some time until it dried up—transferring dried roti to a direct gas flame until it puffs up.

Ingredients
Rheology
Kneading process
Soak@ambient temp

Pressure! Angle!
Direction
Stroke Limit
Snap off

Pre heat
Soak temp vs Time
@ controlling flame

Ramp rate
Peak temperature

The following are necessary for successful Roti puff-up techniques: 1. The type of ingredients and their mix ratio; 2. Kneading the Dough and keeping it aside for some time; 3. Rolling up with an angle and with some pressure; 4. Cooking the roti on a heated Tawa until it dries (soaking time and soak temp); applying high heat(Reflow temp) to puff it up, and then cooling it systematically.

If you transform each process in the above example into engineering, that technical process is called Reflow profiling. For good soldering, we precisely follow the same processes as homemakers follow. In our process, the ingredients can be metals with some proportions, these ingredients mix with the flux and knead as well. Bringing this solder paste to ambient temperature before Stencil print application, setting the reflow temperature, setting preheat temperatures to activate flux, drying the flux, finally raising it to the highest temperature for a few seconds, and finally cooling down the PCBA. For a good solderability reflow process, steps need to be optimized (temperature Vs time) according to the rheology of the solder paste.

INTRODUCTION - REFLOW PROFILE

A profile is a function of temperature applied to the PCB assembly over some time. For a perfect solder joint, the populated PCB must undergo different phases in reflow such as pre-heat zone, Soak zone, reflow zone, and cooling zone. Profile guidelines are normally recommended by the solder paste manufacturer according to the characteristics of the solder paste. Based on set guidelines, setting the conveyor speed and zone temperatures and knowing the heat transfer from the oven to the populated assembly is the keys to success for profiling.

This heat transfer is determined by the PCB size, PCB thickness, PCB substrate, layout, and components–density, mass, size, package, material, and their thermal resistance. Hence, the process window varies for different product assemblies.

High-quality and low-defect soldering requires identifying the optimum temperature for reflowing the solder paste and populated board. Optimum temperature means the best possible temperature at those reflow zones, considering the specified conveyor speed, which results in the best flux management system.

OBJECTIVES OF PROFILING

Reflow involves heating the assembly of boards, where components and circuits get interconnected, bonding through solder paste. The main objective of reflow profiling

- To remove flux volatile

- To start flux activation

- Bring the materials being joined to a temperature that is sufficiently high for a final 'spike' of heat to perform the soldering operation.

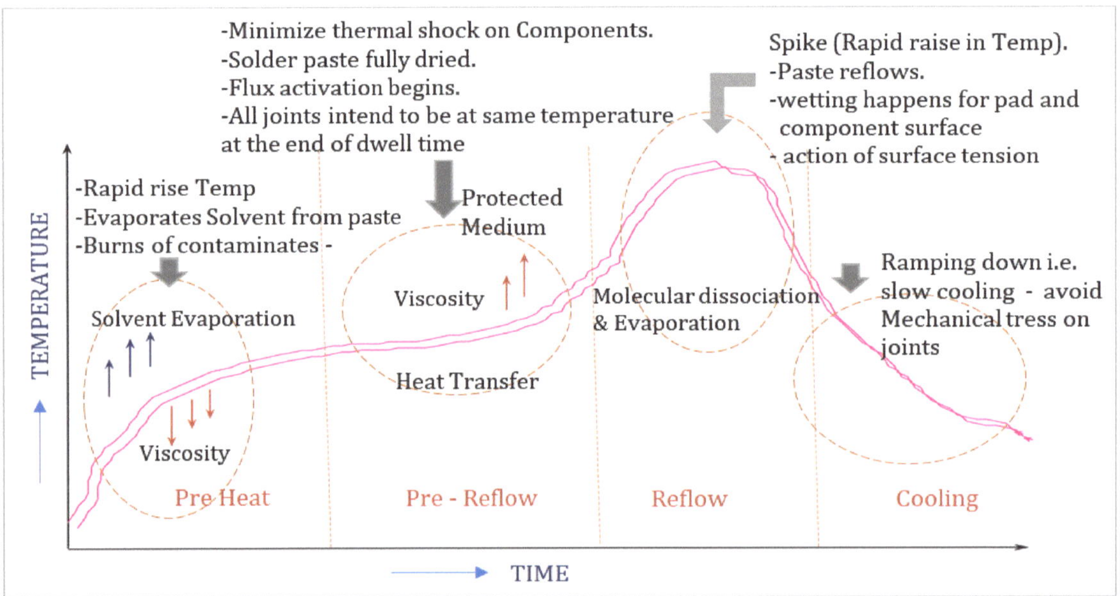

- Bring the materials being joined to a temperature that is sufficiently consistent for the solder to flow evenly onto all the surfaces.

- To reflow the solder, paste to form a liquid solder volume enough to create a sound joint.

- Finally, assembly is to cool to solidify the solder joints

PRE HEAT

The main objective is to minimize the thermal shock to the PCB assembly. Preheat an assembly to a temperature between 100°C and 150°C. The critical parameter in the preheat section is to control the rate of temperature rise between 1°C - 4°C /sec. Preferably less than 2°C is recommended. A ramp rate of 0.5°C – 2°C /sec is normal and can be set by the conveyor speed. As per the experiment, the average ramp rate is 0.78°C /sec.is the best. The set temperature (ΔT) between the heating Zones that fall under preheat is normally < 40°C

Pre-Reflow / Soak Zone

The objective is to ensure the solder paste is fully dry before hitting the Reflow zone. The soak temperature is controlled in a tight range within the specified Time. It is characterized by a consistent temperature of 150°C —170°C for 60 -90 seconds. The maximum recommended duration time is 120 seconds.

Thermal stabilization for large and Small Components happens in this zone to ensure uniform heating of the board Assembly before entering the reflow Zone. The profile allows the thermal Gradient across the PCB assembly. In the case of convection reflow Machines, the need for thermal stabilization is reduced as the entire profile tends to be uniform.

Solder paste selection determines the profile. Different profiles apply to different pastes. To identify optimum temperatures, several trials with permutations and combinations of each zone temperature and the conveyor speed are needed.

Reflow Zone

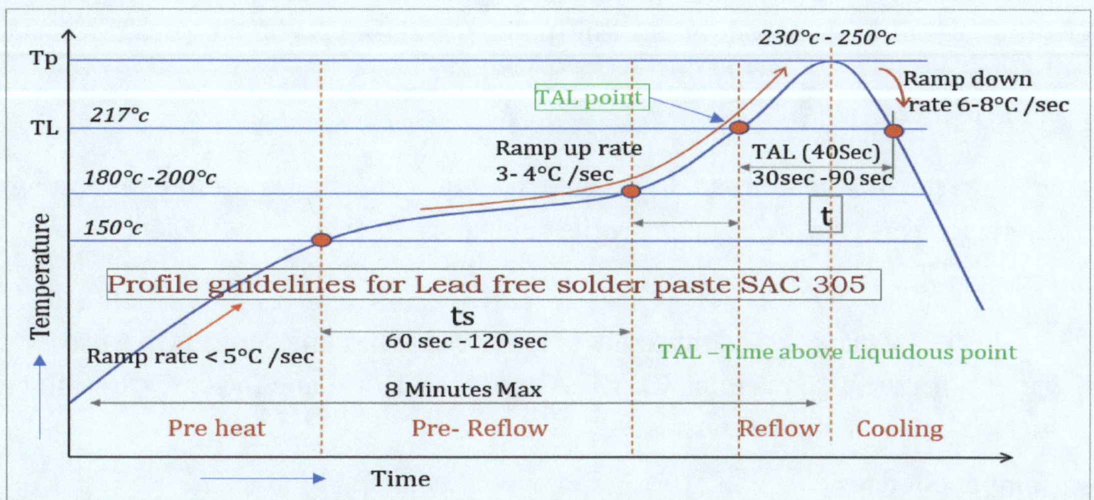

Solder alloy involves the creation of mechanical and electrical bonds through the formation of Tin and copper intermetallic. In forming optimum intermetallic, two critical parameters are involved

in the reflow phase: Peak Temperature and TAL (time-above-liquidus). The peak temperature is generally 20-30°C above the liquidus temperature of the alloy, and the TAL is typically 30-90 seconds to form an effective intermetallic. TAL also refers to wetting time.

The reflow zone profile elevates the solder paste to a temperature greater than its melting point. For Sn63/Pb37 solder, the melting temperature is 183°C, whereas for lead-free (e.g. SAC 305) the melting temperature is 217°C. This liquidus temperature must be exceeded by approximately 20°C for leaded paste, whereas for Lead-free 38°c -48°c to ensure quality reflow for every solder joint lead.

Cooling

The cooling phase determines the grain structure of the solder joint. A fine grain structure provides the most reliable mechanical bond. To achieve this structure, a rapid cooling rate as the solder transitions from liquid (liquidus) to solid (solidus) is needed (the first ~50°C of cooling). A cooling rate of ~4°C-6°C/second is normal.

Table Pb Free Process – Pre-Temperature Tp			
Package Thickness	Volume mm³ < 350	Volume mm³ 350 - 2000	Volume mm³ > 2000
<1.6 mm	260°C	260°C	260°C
1.6mm To 2.55 mm	260°C	250°C	245°C
>2.5mm	250°C	245°C	245°C

Once optimum temperatures are identified, where we can have the best-quality solder joints, that data is called a profile. Temperature profile data are unique for each product or the same family of products. The same profile may not apply to all populated printed circuit boards.

Profile Considerations

Before creating a profile for any product, the following self-explanatory parameters must be considered.

- Solder paste characteristics -

  What type of solder paste is defined for the product for reflow soldering? Paste manufacturers provide suitable profile guidelines for that type of solder paste.

- PCB board -

  The PCB Base material and thickness of PCB need to be considered as thermal expansion varies with the base material. Ex, Glass epoxy, FR4, etc., and in some cases, the type of surface finish should be considered.

- Component density –

  Heat transfer should be uniform even for high-density populated PCBs.

- Component datasheet –

Have details of the components. The lowest temperature at which components can withstand and the maximum temperature. i.e., upper-temperature limit of components. Some components may tolerate the set peak temperature for the profile.

- Lead-free Soldering? –

  This is the most important factor as a lead-free profile requires higher liquidous temperatures.

- Air or inert atmosphere

- Machine capability–

The controlled temperature needs to be set according to the number of Reflow zones, the size of each zone, and the length of the conveyor.

EFFECT OF REFLOW PROFILE – LEAD-FREE

Due to the higher temperature requirements of the lead-free reflow process, new thermal profiles are required for a reliable product. In addition to that, the process window for lead-free alloy is considerably narrow because the liquidus temperature has increased from 183°C (eutectic SnPb) to 217°C-221°C (lead-free SnAgCu, commonly referred to as SAC alloy). This increase of 34°C - 38°C in liquidous temperature means that the required peak temperature range must also increase. Current lead-free solder pastes require peak temperatures in the range of 230°C - 250°C; these peak temperatures can come very close to the maximum temperature limits of many components.

Board laminates are also often stressed even further with these peak temperatures much higher than the glass transition state of the laminate. Given the higher operating temperatures of lead-free alloys, the heat transfer efficiency also requires strict temperature variation reduction across the entire assembly.

Carefully consider your components' upper-temperature limit against the peak temperatures of lead-free alloys, and you can see the value of the smallest possible thermal differential.

REFLOWING DOUBLE-SIDED PCB

The key issue with double-sided reflow is that the first side of the PCB is reflowed twice, which 'ages' the joints. During the second reflow, the previously wetted joints are reflowed without the benefit of any protective flux. Oxidation reduces the surface tension of the solder, causing dropped components. Chances of Dewetting and Components on the underside during the second reflow operation are supported only by the surface tension of the solder or by the adhesive process. An adhesive process is preferable.

In this case, while making profiling, both primary and secondary reflow profiles should meet the same target specifications. The board assembly has greater thermal mass during the second reflow. Hence, different profiles are required for primary and secondary.

REFLOWING PCB IN DUAL CONVEYOR REFLOW

Dual conveyor Reflow Oven

Reflow profiling may not be very complex if you are running the Top and Bottom PCB of the same PCB or the same family of PCBs in a dual line reflow oven. Profiling is quite complex for different families of PCBs running simultaneously in dual-lane, single-chamber reflow ovens. Profile optimization for two different PCBs with different conveyor speeds is quite complex.

For this type of process recommendation, the PCB design and the selection of paste type play a critical role in reflowing double-sided PCBs in dual-lane, single-chamber reflow Ovens. Solder pastes need to be selected to ensure that flux residues have some continuing protective capability for second-side reflow. Also, they should minimize TAL (time above liquidus) to remove any sources of shock or vibration.

One strategy normally considered is using an inert soldering atmosphere to reduce oxidation and the demands made of the flux and increase the process window. Another approach is to go for the latest models' dual-lane dual-temperature ovens (dual chamber), which allow two different thermal profiles to run simultaneously in the same reflow oven footprint.

SETTING REFLOW ZONE TEMPERATURES

Individual zone temperatures can be set at approximately. to initiate the profiling process for any given product. However, one needs to have complete knowledge of the reflow profiling process and details of the Machine configuration and the solder paste.

The first thing to consider is your solder paste manufacturer's recommended profiles for the paste formulas you will be using. Solder paste manufacturers generally provide a wide window (in terms of total heating time) for the various stages of the reflow profile: 150 to 240 seconds for preheat and soak, and 30 to 120 seconds for reflow/time above liquidus.

GUIDELINES TO SET ZONE TEMPERATURES

Example

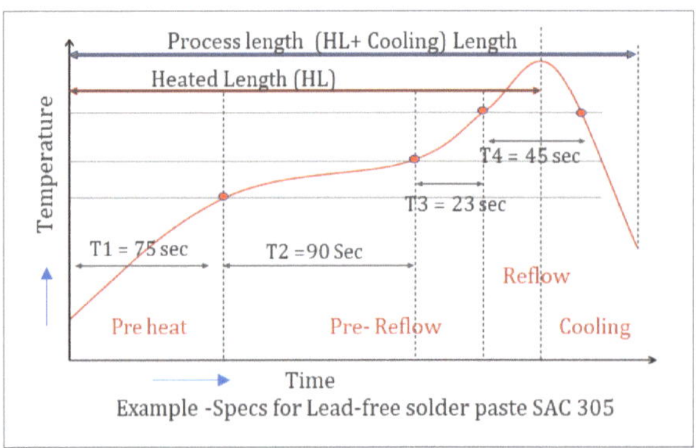

Example -Specs for Lead-free solder paste SAC 305

Production requirement

- 8inch PCB length needs to be processed

- 180 Boards/hour approx. Target

- Selected solder pastes SAC305

Let us consider the solder paste specification recommended by the paste manufacturer's 4 Min three-step profile.

- Step1:

- Boards / min = 180/60 = 3 boards/min

- Length of the board = 8 inches

$$\text{Line speed} = \frac{(Boards\ /\ X\ (Board\ length\ )}{Load\ factor}$$

a) Use Formula

$$\text{Load factor} = \frac{Length\ Of\ board}{length\ of\ Board + Space\ between\ boards}$$

$$\therefore \text{Load factor} = \frac{8\ Inch}{8\ inch + 2\ inch} = \frac{8}{10} = 0.8$$

1. Process dwell time = 4 Min

1. Use Formula

$$\therefore \frac{3\ boards\ /}{0.8}$$

$\therefore$ Line speed (Conveyor speed) = 30 Inches / Min

$\therefore$ Reflow must have a processing speed of 30"/min

Use the formula

$$Process\ speed = \frac{Oven\ chamber\ heated\ lentgth}{Process\ dwell\ time}$$

$\therefore$ 30 in/min = (Reflow heated length)/ (4 min)

$\therefore$ Reflow Heated length = 30 X 4 = 120 inches

Preheat 30°C (room temperature) - 150°C Max rise slope / Ramp Rate 1 to 2 °C = 75Sec (T1)

- Soak / Dwell 150°C-180°C for 60 to 120 sec = 90 Sec (T2)

- Ramp from 180°C to 217°C = 23 sec (T3)

- Time Above Temp 245°C for 30-60 sec = 45 sec (T4)

Step 2

$\therefore$ Total Time(T1+T2+T3+T4) = 233 sec

∴ Total Time in min = 3.9 min @ Temp

1. Use the formula

$$Conveyor\ speed(inch/min) = \frac{Process\ length}{Time\ required\ in\ Oven}$$

Step 3

Required Belt speed of the oven 3.9 minutes @ Process

1. Process length (PL) = Heated Length (HL) + Cooling 12" to 20".

A Cooling Zone is added because the process is not complete till the solder joint is solidified.

∴ process length (PL)= 120" HL + 20" = 140"

Step 4

Zone Size and Zone Time Required:

1. The heated length is 120" / 11 zones = 10.91" per zone

2. Conveyor speed = 30 inch / min = 2 inch/sec

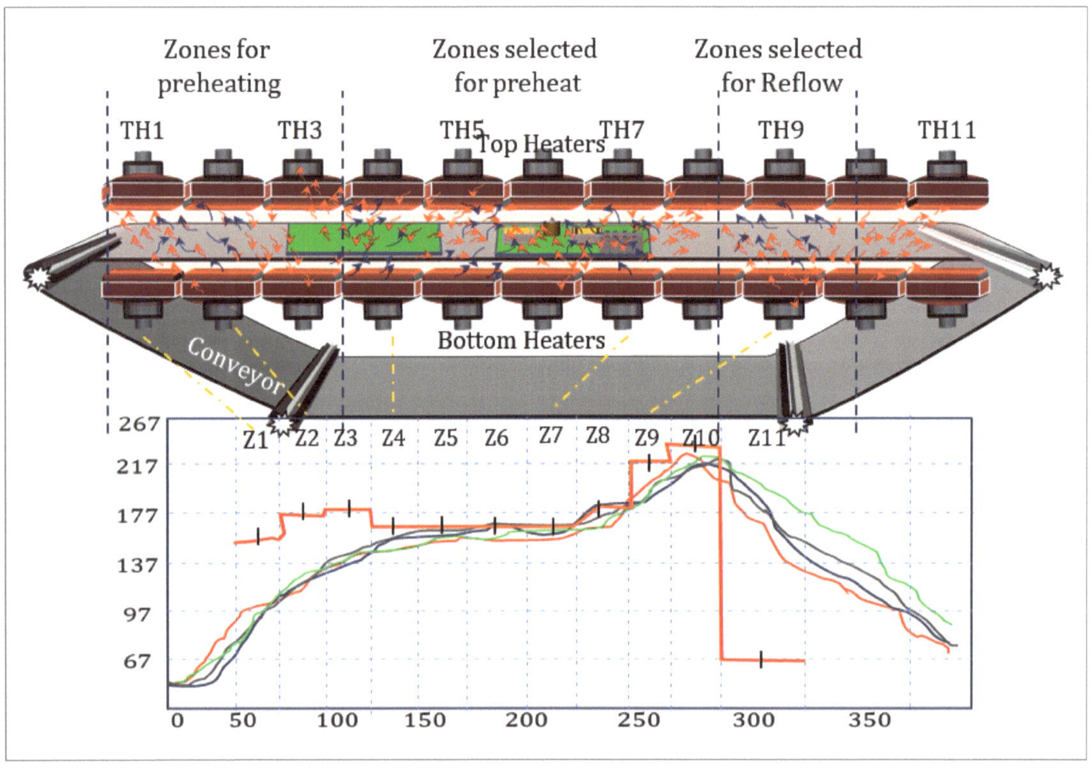

At 10.91" per zone, the PCB will stay in each zone for 21.82 sec i.e. 22 seconds

∴ for T1, 75 sec = 75/22 = 3.4 zones

∴ for T2, 90 sec = 90/22 = 4 zones

∴ for T3, 23 sec = 1 zone

∴ for T4, 45 Sec = 45/22 = 2 Zones

Step 5

Setting zone Temperatures:

- Given that Convection is 90% efficient, choose the zone temperatures you know as the board leaves the zone, and multiply by 110%.

- 150*110= 159°C Z4

- 190*110%=210°C Z8

- 220*110%=242°C (Zone9)

- 245*110%=270°C (Zone 11)(Peak temp)

Then just fill in the blanks for the Remainder of the zones. Space the temperature changes evenly.

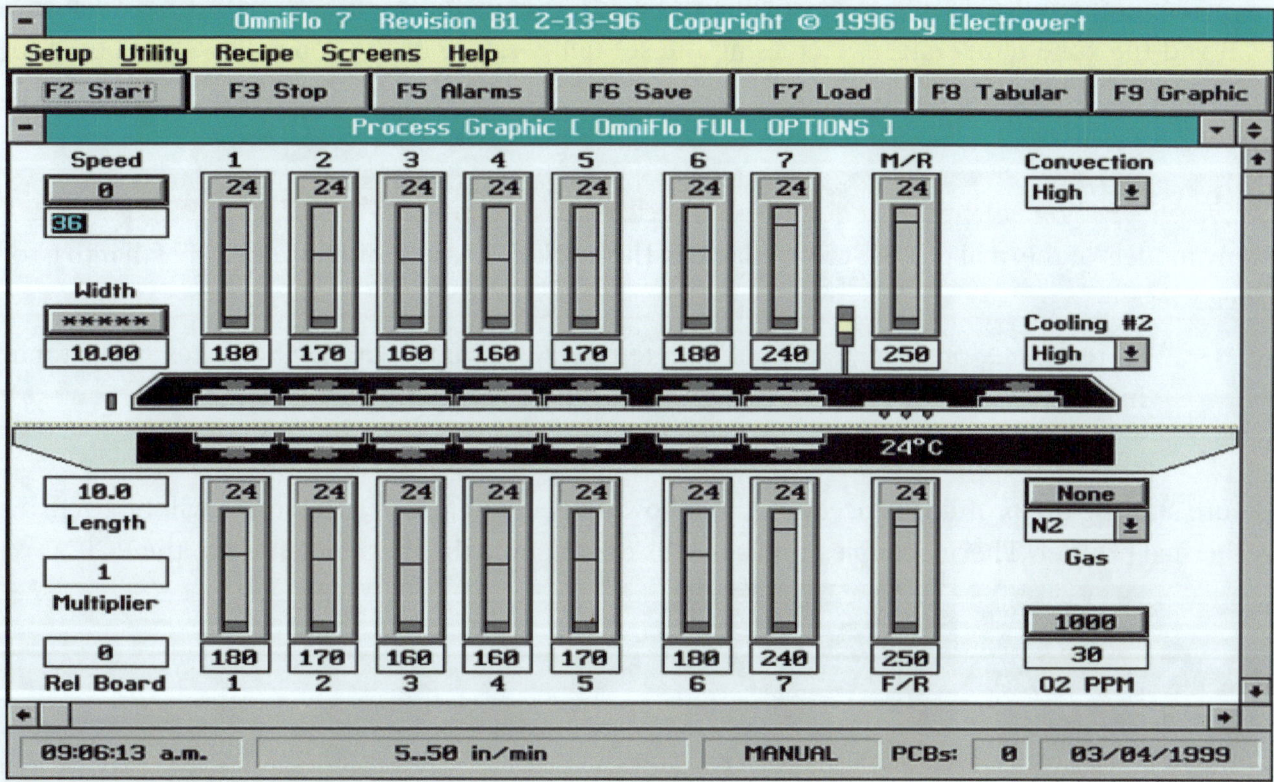

REFLOW PROFILING A PCB ASSEMBLY

For getting the typical or recommended/ calculated reflow profile, we have a choice only on two key parameters, i.e., conveyor speed and setting temperature at each zone. The number of the Zones in the machine is fixed. The zone width is fixed.

To ensure convection efficiency towards PCB assembly while reflowing, we need thermal instruments, often called Trackers or profilers, to measure the actual heat gradient characteristics towards high-mass components, low-mass components, and PCB surface area in the reflow process at each point of time and under specific zones. This data helps to set the optimum profile for each assembly.

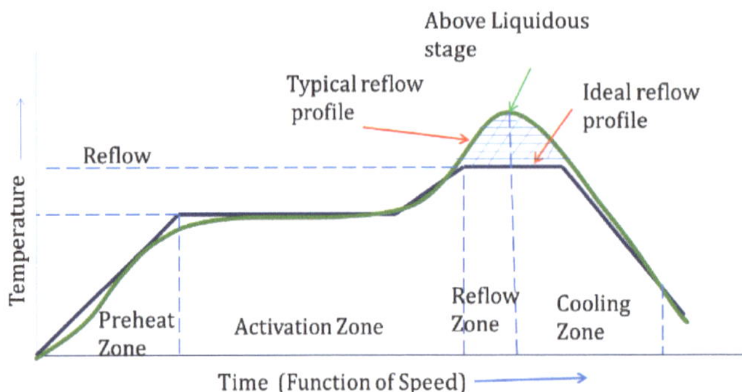

When graphed, a Typical reflow profile curve forms that represents the required temperature at a specific point on PCB at a given point of time throughout the reflow process. Belt speed determines the duration at which the board is exposed to a specific set temperature at each zone.

The temperature setting in each zone produces a temperature differential Δt between the PCB and the zone temperature. Increasing the set temperature of the zone allows the board to reach a given temperature faster.

PROFILE - GETTING STARTED

Many models of thermal profiles are available in the market. These profilers generally fall into two categories:

- A real-time profiler transmits the temperature/time data and creates the graph instantaneously.

- Other profilers capture and store the information for uploading to a computer later.

Before starting the profiling procedure, the following equipment and accessories will be needed: A thermal profiler, Thermocouples, and a means of attaching the thermocouples to the PCB.

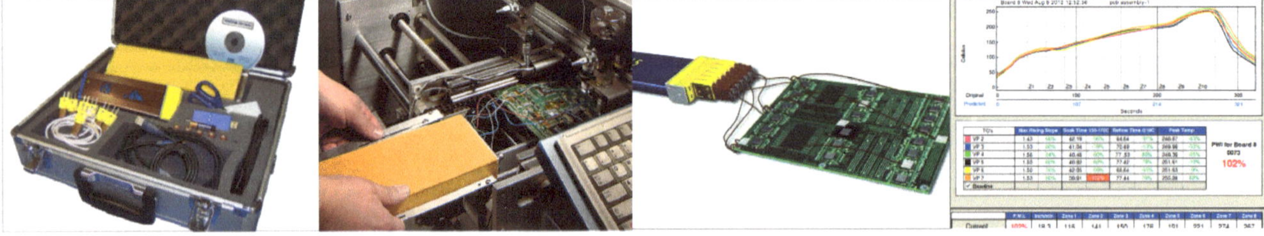

A profiling accessory kit contains the necessary jigs to carry the profiler inside the reflow process. Normally, the profiler supplier supplies this kit.

Thermocouples that are used to profile a PCB should be of sufficient length as required for the profiler and must be capable of withstanding typical oven temperatures. In general, thinner gauge thermocouples are more desirable because they produce accurate results - smaller heat mass increases responsiveness.

THERMO COUPLE

A thermocouple is a temperature measurement sensor containing two dissimilar metals joined at one end (a junction) that produces a small thermoelectric voltage when the junction is heated.

If two wires of different materials are joined at their ends and one end is maintained at a higher temperature than the other, a voltage difference will arise, and an electric current will exist between the hot and the cold junctions. This phenomenon can be utilized for the accurate measurement of temperature using a thermocouple in which one wire junction is maintained at a known reference temperature and the other at the location where the temperature is to be measured.

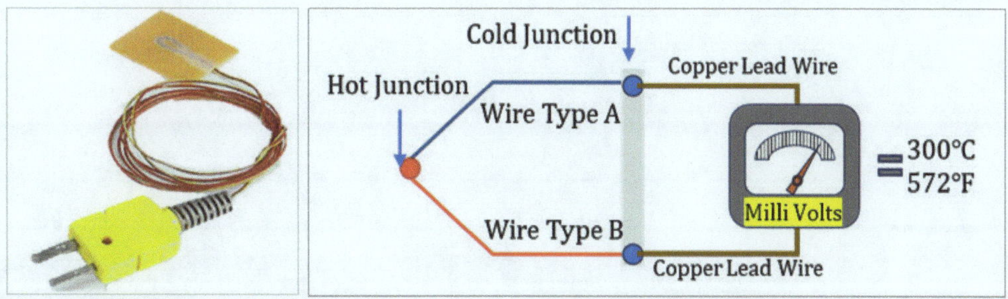

The most common thermocouple used in the electronic industry is the K type. It is made up of Chromel (Ni-Cr) configured as Positive (+), and another wire of Alumel (Ni-Al), referred to as Negative (-).

The maximum temperature range for a K-type thermocouple is 270°C to 1372°C. The smaller the wire diameter, the faster the response time. Recommended 36AWG (0.005inchor 0.127mm)

ATTACHING THERMOCOUPLE TO PCB - METHODS

There are several ways to attach thermocouples to the PCB.

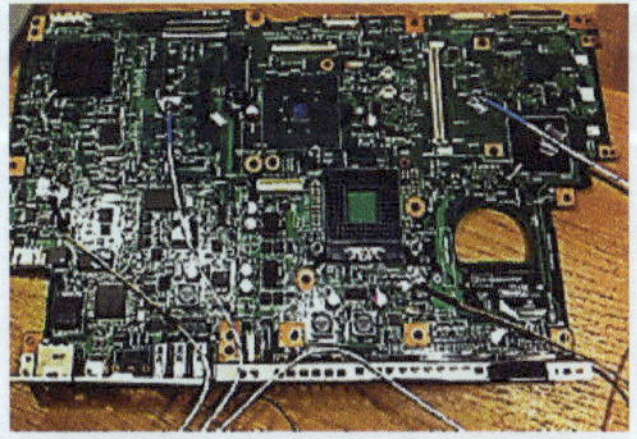

Method 1:

Attach thermocouple tip using high-temperature solder such as silver or tin alloy. This is the most preferred method. Use the smallest amount of solder.

Method 2:

It is a non-destructive method. A small amount of thermal compound (also known as thermal conductive cream or thermal grease) is applied to the thermocouple tip, which is then attached using high-temperature Tape, Kapton tape.

Method 3:

Attach the thermocouple using a high-temperature adhesive such as cyanoacrylate. This method is not suitable for PCBs.

GUIDELINE TO ATTACH THERMOCOUPLE (TC)

It is best to attach the thermocouple tip between a PCB pad and the corresponding component lead.

- High Temperature solder

TC (Thermocouple)attached to the Solder Joint and on the metal Parts

- High-temperature epoxy and Adhesive

TC attached to the Component body and TC attached to the Glass Epoxy

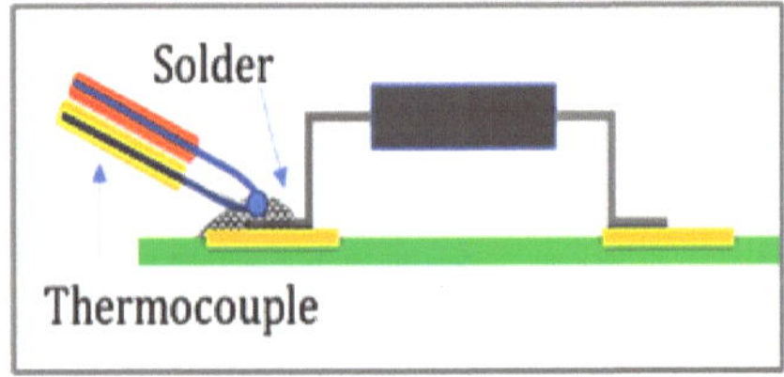

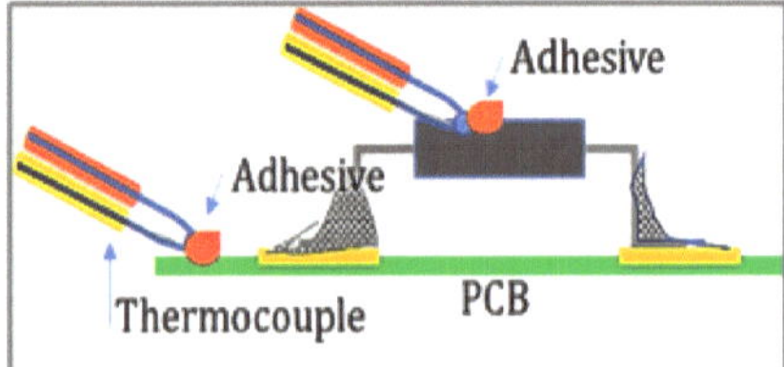

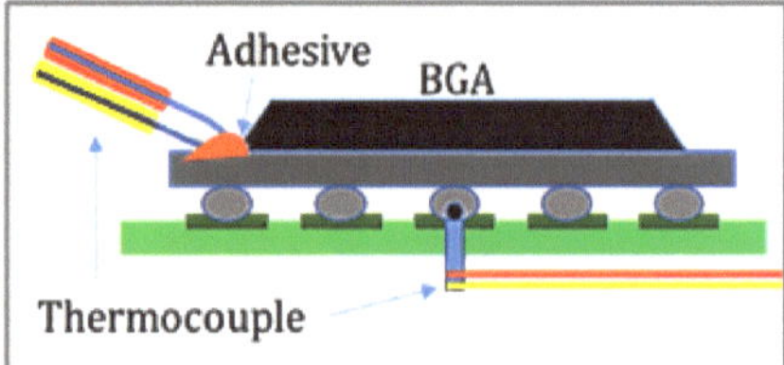

- TC attachment for BGA

Drill hole through the pad - attach BGA-drill through the Ball - fill with Epoxy - insert a thermocouple

Up to six thermocouples or even more may be placed on the assembly, and proper selection of their locations is important. The number of TC requirements depends on the size of the PCB and the density of the components. Unless there are specific known problem areas, the thermocouples should be placed in areas to provide

- maximum peak temperature,

- maximum rate of temperature change

- maximum and minimum dwell above wetting temperature.

The maximum peak temperature will normally occur along the edges of the PCB near low-mass components such as chip resistors or capacitors. The minimum peak temperatures will normally occur near the center of the PCB near high-mass components such as large IC packages.

SAFETY FACTORS

- Care should be taken that the trailing thermocouple bundle does not become entangled with the conveyor mechanism.

- The thermocouple leads should not have tension. Make sure that the feeding of the thermocouple through the reflow should not be slowed down in between.

- The thermocouple polarity needs to be maintained

- The thermocouple crossed wire is not acceptable when soldering or attached to the Substrate.

- Grounding is required for the profiler to reduce electric noise.

POST PROFILE CORRECTIVE ACTIONS

The key objective of reflow profiling is to achieve reliable solder joints at every point in PCB assembly without deteriorating the substrate properties of packages and printed circuit boards. This means defect-free solderability.

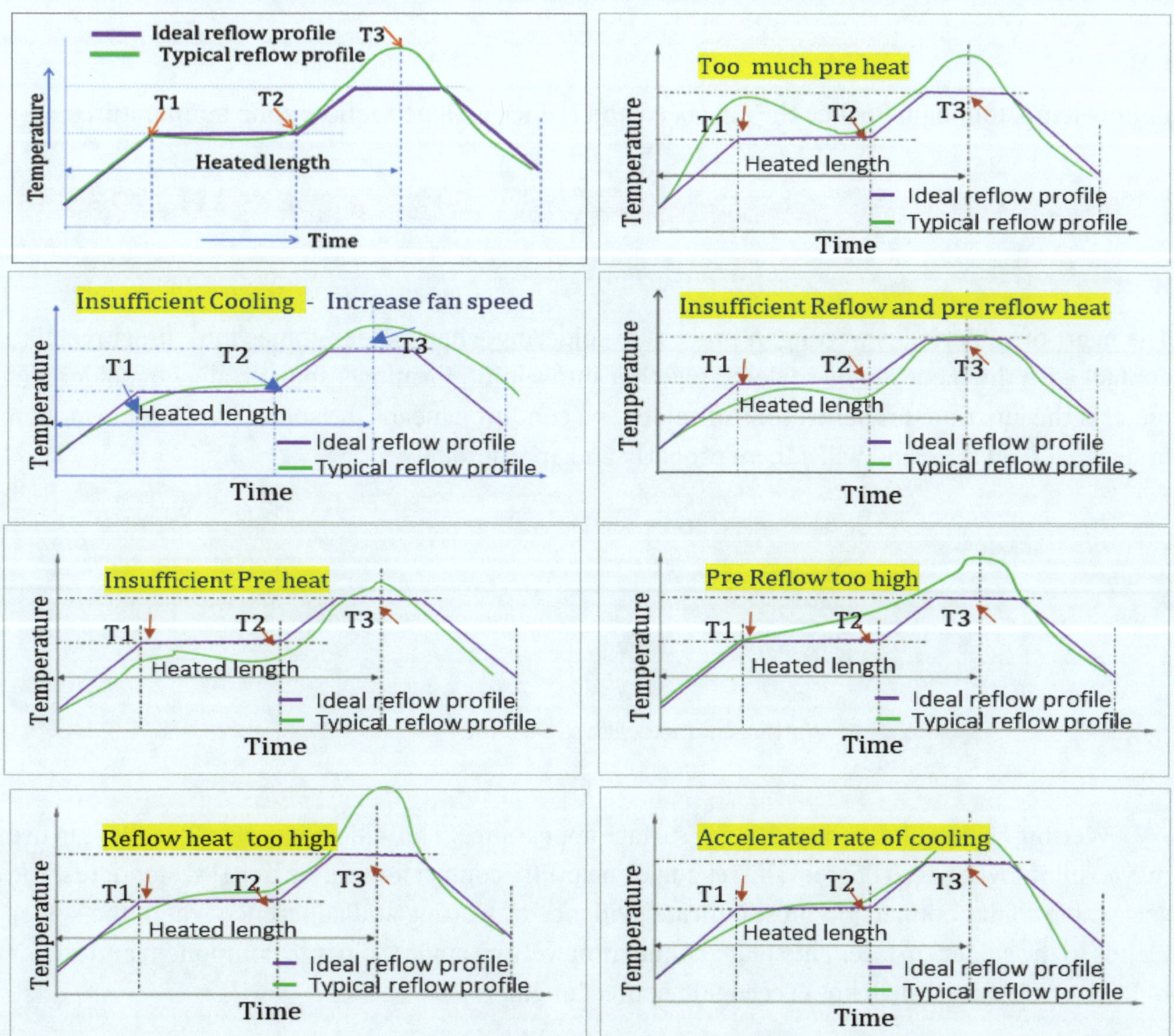

Let's consider a Typical reflow profile for a particular solder paste with reference to the Ideal reflow profile recommended by the paste manufacturer. As per the production target and using guidelines, calculate and set Reflow zone-wise temperatures and conveyor speed.

To optimize the profile, one has to perform several profile trials with the tracker and corrective actions based on profile results. Here are some examples of profile results and proposed corrective actions.

Case 1

Insufficient pre-heat. Need to increase oven zone temperatures assigned for preheating.

Case 2

PreReflow Temp is too high. ramp-up rate needs to be less. Reduce the Zone temp assigned for pre-flow.

Case 3

Reflow temp. too high. Reduce the fan speed and reduce assigned reflow zone temperatures.

REFLOW SOLDER DEFECTS

Wetting

The heart of soldering is wetting. A process in soldering where solder comes into direct metallic contact with the metals to be soldered together into a joint, forming a specific alloy of solder and metal at the junction. Proper wetting of solder as a condition means the solder has become molten in its ideal fluid state and will adhere properly and appropriately.

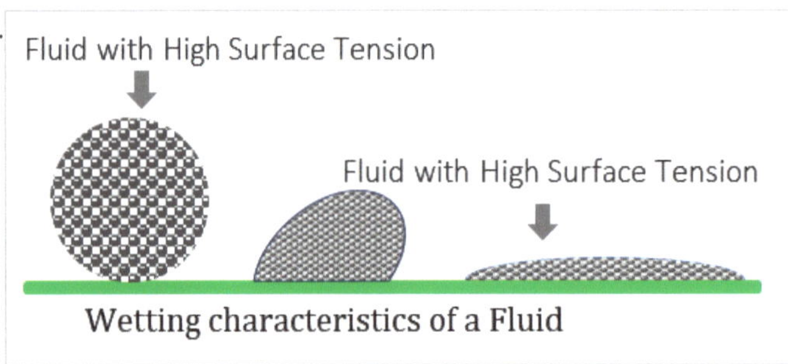

Wetting characteristics of a Fluid

Wetting cannot always be judged by surface appearance. The wide range of solder alloys in use may exhibit low or zero degree contact angles near 90° contact angles. As per IPC standards, the acceptable solder connection must indicate evidence of wetting and adherence where the solder blends to the solder surface. The solder connection wetting angle (solder to component and Solder to PCB termination) shall not exceed 90° contact angle.

SOLDER ANOMALIES

Many factors contribute to solder defects. The origin of these defects can be from paste printing, component placement, process handling, the reflow process, and even production environment conditions. Finally, the key objective of solder technology is to reduce solder defects. Hence, optimizing the solder reflow process considering the changes in printing parameters, printing process, and Placement process is a big challenge.

Before analyzing the solder anomalies, one should know what a good solder joint is and what the target and acceptable requirements. Here are some examples of possible solder anomalies related to reflow profiling.

NONWETTING

IPC –T-50 standard defines Nonwetting as the inability of molten solder to form a metallic bond with the base metal. Visible base pad. Difficult to solder. Hence the possibility of non-wetting.

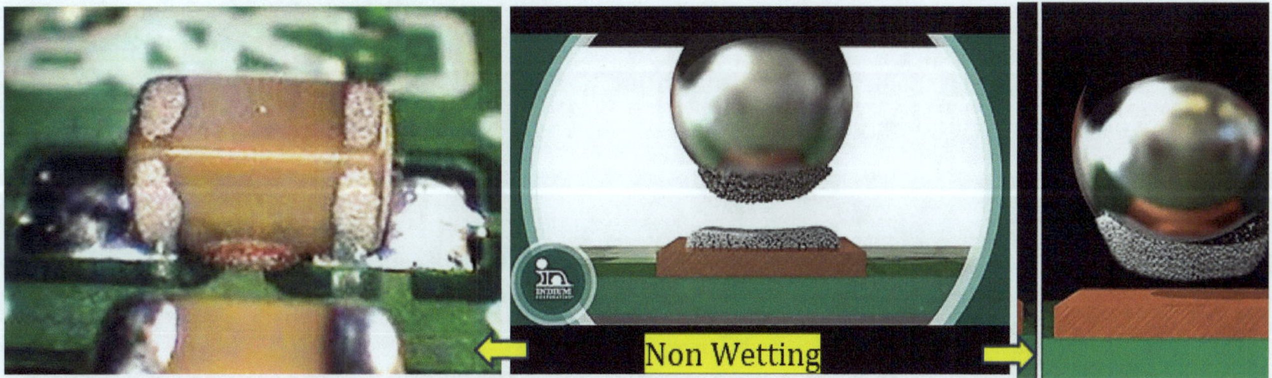

Possible cause owed to reflow - Insufficient heat during the reflow process where the flux is not activated properly. Reflow profile settings need to be analyzed considering the solder paste Manufacturer's datasheet.

COLD SOLDER JOINT /DULL JOINT

IPC –T -50 defines a cold solder connection as " a solder connection that exhibits poor wetting and that is characterized by a grayish, porous appearance.

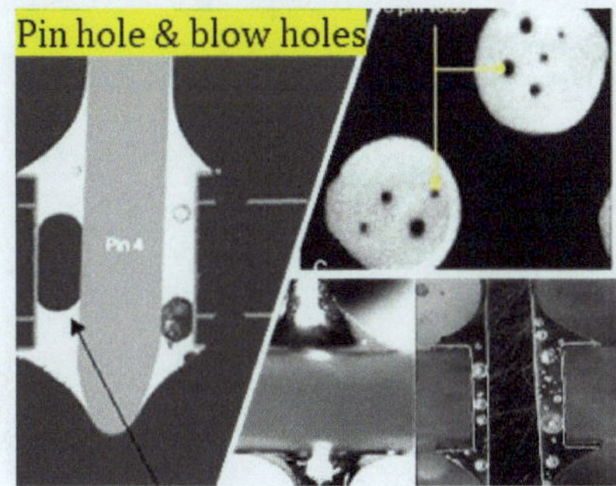

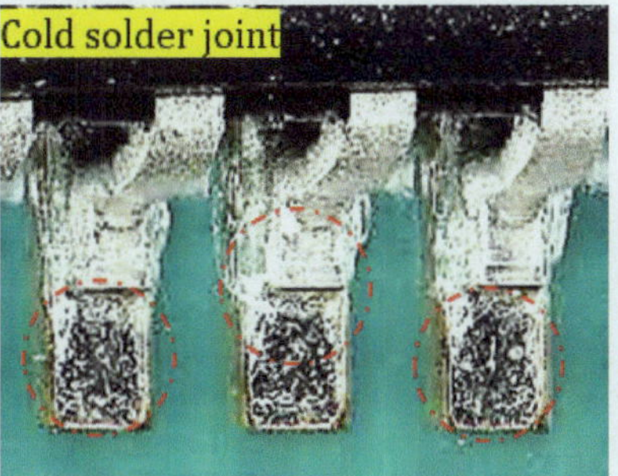

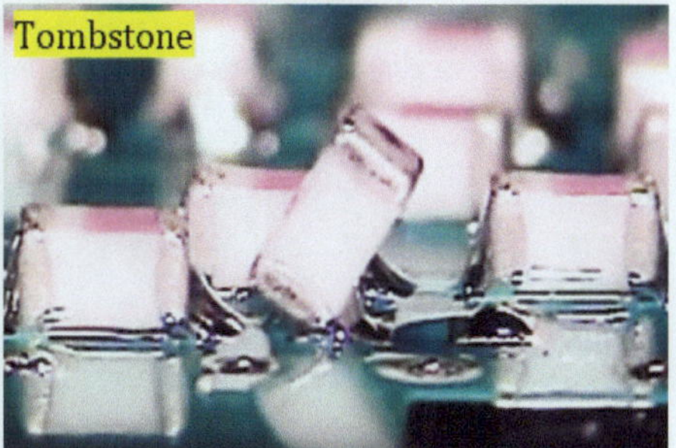

Possible causes owed to reflow- Insufficient heat present to reflow the solder adequately. Another cause can be accelerating the cool-down rate after reflow.

Excess solder - Solder Balls

Solder balls are Spheres of solder that remain after the soldering process. Solder balls create an artificial bridge between two adjacent leads, causing functional problems to an electrical circuit. Possible causes owed to reflow- Improper reflow profile can also cause solder balling.

Voids

Voids are cavities or gas pockets inside the solder joint formed by gases that are released during reflow or by flux residues that fail to escape from the solder before it solidifies. Voids reduce the solder connection below the minimum acceptable

Possible causes owed to reflow, Preheat too aggressively for flux. Improper profile for solder paste

Pin Holes and Blow Holes

Blow holes or pin holes are basically the same thing. Gases entrapped inside the solder joint try to escape during soldering and leave holes.

Tomb Stoning

Tomb Stoning occurs when -Solder paste approaches reflow temperature - Solder begins to wet unevenly - Surface tension increases on the first side of the part to reflow - The part is pulled free of the second pad before a metallurgical bond can be achieved.

Possible causes owed to reflow, profile experiments not done accurately.

SUMMARY

This technical content includes detailed descriptions of Electronic manufacturing processes and the manufacturing enterprise that can help aspirants, students, and readers of these articles understand "a little about everything". The relevant content of this report is based on the author's experience with Speedline Reflow machines, IPC 610F, material absorbed from various articles, white papers, and technical documents that are published/ released and uploaded by famous scientists, SMT professionals, and design engineers of OEMs.

Year 2000
Organic acid-based solder paste
Rosin Based
Year 2014
Lead free world
Low solids , No clean fluxes
Flux management system
Wider process window for critical reflow parameters
SMT realized growing importance of thermal profile
TAL
Soak PT
New heating Technology Convection
Inert nitrogen introduced as an alternate to air
Up until this time, IR heating was the choice of heat transfer medium.

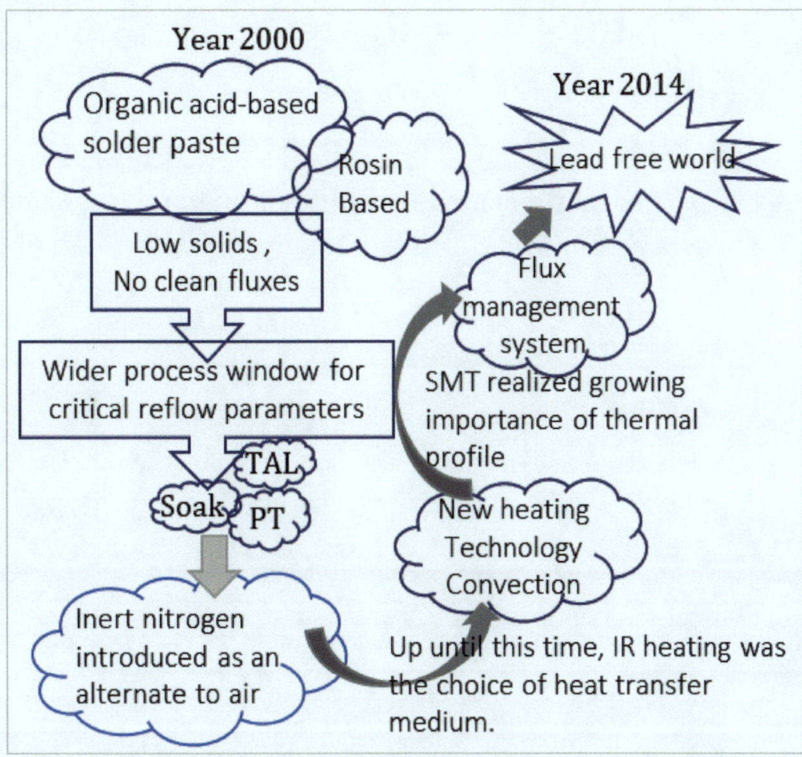

End Of Part I

www.ingramcontent.com/pod-product-compliance
Lightning Source LLC
Chambersburg PA
CBHW041640110726
48005CB00003B/661